MUSEUM OF Science™
ACTIVITIES FOR KIDS

MUSEUM OF Science™
ACTIVITIES FOR KIDS

TANYA GREGOIRE
WITH
JOAN PARISI WILCOX
ILLUSTRATED BY
JULIE FRAENKEL

Adams Media Corporation
Holbrook, Massachusetts

Published by Adams Media Corporation
260 Center Street, Holbrook, MA 02343

ISBN: 1-55850-633-0
Printed in Canada

First Edition
J I H G F E D C B A

Library of Congress Cataloging-in-Publication Data
Gregoire, Tanya.
Museum of science activities for kids / developed
by The Museum of Science in Boston; written by
Tanya Gregoire with Joan Parisi Wilcox;
illustrated by Julie Fraenkel; photographs by Eric
Workman.
     p.   cm.
   ISBN 1-55850-633-0
1. Science—Experiments. I. Wilcox, Joan Parisi.
II. Title
Q164.G745 1996
507'.8—dc20   96-28023
                           CIP

Museum of Science Staff:
Belinda Recio, Manager of Book and Product
Development
Sally Ellison, Research and Illustration
Coordinator

Science Reviewers:
Maureen McConnell
Paul Ellingwood

This book is available at quantity discounts for
bulk purchases.
For information, call 1-800-872-5627
(in Massachusetts, 617-767-8100).

Visit our home page at:
http://www.adamsmedia.com

DEDICATION

To my husband John, whose love and support have provided me with a steady source of strength, and whose interest in my experiments is beyond the call of duty (especially the spider hatchlings in the kitchen). And to my family, friends, and colleagues (many at the Museum of Science) who have joined me on my journey through lifelong learning.

TABLE OF CONTENTS

CHAPTER 1

SUN, MOON, & STARS

CONSTELLATIONS
MOON JOURNAL
SUN CLOCK

INTRODUCTION

Once upon a time there was nothing. No people, no houses, no cars. No animals, no trees, no rocks. There was not even an Earth, Sun, or solar system. Believe it or not, once upon a time there was nothing at all. Okay, there was one teeny tiny thing —an infinitesimally small point of energy. At least that's one theory. Some scientists think of this tiny bundle of pure energy as the "primeval cosmic egg." But it was much smaller than an actual chicken egg. In fact, it was so small that it would have made the head of a pin look big! Until one day, about 15 billion years ago, it exploded. Bang! Actually, scientists call this explosion the "big bang." When this tiny ball of energy exploded, it began a chain reaction of creation. From that tiny explosive beginning came everything we call the universe. Including you!

Of course, nobody witnessed the big bang, so scientists cannot say for sure that it happened. Therefore, the big bang is just a theory—a general explanation based on scientists' ideas and observations, but not proven for sure. Many scientists believe the big bang theory because they have discovered clues to its occurrence by studying

the properties and movements of heavenly bodies such as planets, stars, and galaxies. All these astronomical bodies seem to be rushing apart, as if they were still feeling the forceful outward push from that first explosion.

The big unanswered question is, What will happen when the energy rush from the explosion finally stops? Some scientists think the movement of the stars and other astronomical bodies will slow down, stop, and finally begin to contract back in toward the center of the universe. One day, perhaps several billion years from now, these collapsing planets and stars will smash into each other, causing the "big crunch." The result will be another huge explosion that starts the outward rush all over again! However, not all scientists agree with this theory. Some think the stars and galaxies and everything else in the universe will keep expanding forever.

One thing is pretty much agreed upon, even though it, too, is still only a theory: The universe has no edges, no real boundaries. Thus, you can never get to the end of the universe. Think of a balloon. If you were an ant walking over the surface of a balloon, you would never reach an edge; you would just keep walking round and round forever. As the balloon expands, its surface gets bigger and bigger, but there are still no edges.

The universe is mostly empty space, even though when we look up at the night sky or through a telescope it seems as if it is filled with stars. Each star is formed in a large cloud of gas and dust called a *nebula*. The gas and dust in the nebula cluster together, and as they get packed together more and more densely, gravity also increases. Gravity is the "attractive" force that pulls the particles of gas and dust together. This force increases as the matter becomes more closely packed.

The densely packed molecules of dust and gas begin to heat up and, being so close together, bang into each other, releasing even more heat. Pretty soon (actually, it takes millions of years) the nebula becomes so hot and dense that the hydrogen atoms get squashed together to form a new element, helium. This process—a nuclear reaction called *fusion*—releases lots of energy in the form of light and heat. And a star is born! Scientists consider our Sun to be an average-size, run-of-the mill star. There are stars believed to be

one hundred times more massive than our Sun! Stars live for millions, sometimes billions, of years. But eventually they use up all their fuel and burn themselves out.

The astronomical body we are most familiar with is the planet—after all, we live on one! But just what is a planet? A clue comes from the word itself: "Planet" comes from the Greek word *planasthai,* which means "to wander." Planets do not wander aimlessly through space, however. They are natural bodies that circle, or *orbit,* around a sun. There are nine planets in our solar system (in order of distance from the Sun): Mercury, Venus, Earth, Mars, Jupiter, Saturn, Uranus, Neptune, and Pluto. Mercury, being closest, takes only a quarter of a year—only three months—to orbit the Sun. Pluto, which is farthest from the Sun, takes 248 years to complete its orbit. Earth, of course, takes one year. The time of our orbit around the Sun is how we determine the length of a year.

There is so much to know about the stars and planets and other aspects of the universe that we could go on forever. Instead, use the following activities to go out and explore the night sky yourself.

CONSTELLATIONS

What do you think you would be doing if you were looking at a bear, a horse, a dog, a fish, and a scorpion? Visiting a zoo? Watching a wacky cartoon menagerie? How about looking at the night sky! All of these animals "live" in the sky as patterns of stars that look like familiar, everyday objects or creatures. These star patterns are called *constellations*. Ancient people had nothing to distract them after a hard day's work except the beauty and fascination of the mysterious night sky. After observing the sky night after night, they noticed that the stars formed patterns, and they began to tell stories about the starry people and animals they imagined in the sky. They also noticed that, over time, the positions of individual stars and of the constellations themselves seemed to move slowly from east to west across the sky. They couldn't explain this movement, but they made up explanations in their stories. For instance, one Native American legend says the stars are hungry animals searching the sky for food. Today, we accept these explanations as beautiful and imaginative stories, but we know that the stars are not animals searching for food. Scientists have told us that the stars do not really move; they just appear to move because we ourselves are moving. That is, the planet Earth is spinning on its axis, like a top; and as we move, the stationary stars seem to move instead.

Ancient people also thought that all the stars were points of light lining the inside of the great, invisible globe of the universe. Therefore, they judged that all the stars were the same distance from Earth. We now know that that belief, too, is incorrect. However, the stars are so far away from us that they all do indeed appear to be at the same distance. To see how ancient people could have formed this belief, just think of yourself standing at one end of a football field. On the ground at the other end are two raisins, one a few inches in front of the other. You probably would not be able to tell which raisin was closer to you; in fact, both raisins would look like they were the same distance away. Those ancient people, however, missed an important clue that could have suggested that all stars are not equally distant from Earth: the stars' differing brightnesses. Not all stars shine with the same brightness, and this is a clue to how far away from Earth a star is. Generally speaking, the brighter a star appears, the closer it is to us; the fainter a star appears, the farther away it is. This is not a hard-and-fast rule, because some really massive and very bright stars could shine brighter than other stars but still be farther away from us. Astronomers have studied stars' brightnesses—called their *magnitude*—and assigned each star a number that indicates its place on this scale of magnitude. A star with a low number is very bright, and one with a high number is very faint. For instance, the star Alpha Centauri B has a magnitude of 1, so it is very bright. The faintest magnitude most of us can see is a magnitude 6. However, using the Hubble telescope—the highest powered telescope—scientists have been able to see a star with a magnitude of 25! Actually, this star turned out to be a group of stars, and the group's light is very faint because the stars are extremely far away.

Ancient people may not have known all the neat facts we know about stars, but they knew one thing that remains true today—the really fun part of star watching is identifying the constellations and making up stories about them. Just about anywhere you look in the sky, you can group stars together into very imaginative constellations. One of the most famous constellations in the Northern Hemisphere is the Big Dipper. But over the centuries, not everyone has interpreted this constellation as a dipper (a long-handled cup). People in Great Britain saw a plow! The Micmac people of North America thought this group of stars looked like a bear being chased by hunters. The Micmac story about this constellation says that the hunters finally killed the bear, and it is the bear's blood falling to Earth that colors the autumn leaves red, brown, and orange.

What stories do you see in the star patterns of the night sky?

I-N-S-T-R-U-C-T-I-O-N-S

There are more than a billion, billion stars in the universe! You will not be able to see them all, but typically, with your naked eye, you will be able to see about two thousand from your neighborhood. That's still a lot of stars! To focus on the stars you want to group into a constellation, you can use a star frame. Take a wire coat hanger and stretch it into a square by pulling the bottom down. Then fold the hook over to make a handle, and wrap it with masking tape to protect yourself from any sharp edges. Now the fun begins!

MATERIALS

- pencil
- notebook
- wire coat hanger
- masking tape
- blanket
- clear, starry night
- safe outdoor observation place
- star-gazing buddy

Go outside on a clear evening and find a nice spot to lay your blanket. As you get cozy, allow your eyes to get used to the dark: It usually takes about ten minutes for them to adjust. Use your star frame to choose a part of the sky to study. Move the frame around until you isolate a group of stars within it that looks like a figure or an animal or object. Use your imagination! In your mind, play "connect the dots" with the stars. Once you create a constellation,

draw it in your notebook, actually connecting the stars with lines to draw the object you imagined. Share your constellation and its story with your star-gazing buddy. When you get home, write down the story that goes with your constellation.

OTHER THINGS TO TRY

Go out on the following night. Can you find your constellation again? Bring a star chart with you and see if you can find the "established" constellations in the night sky.

FUN FACTS

- In the Big Dipper there is a double star, which can be used as an eye test! At the bend in the Big Dipper's handle are two stars very close together. The one called Mizar is the easiest to see. But can you spot its dimmer companion, Alcor?
- The ancient Romans believed that everyone had a star in the sky. When you died, your star fell from the sky and disappeared.
- The planet Pluto was discovered in 1930. It was named by an eleven-year-old girl named Venitia Burney, the daughter of a professor of astronomy.

MOON JOURNAL

Forget bedtime! There are too many interesting things to see in the night sky for you to go to bed as soon as the Sun goes down and the Moon comes up. The Moon itself is very interesting. No, not because it is made of green cheese—we know that is just an old fable—but because the Moon seems to change shape. It is one of the few objects in the night sky that seems to look different at different times. One night it is fat and round and full, and a few weeks later it appears to be cut in half, and still later it is just a sliver of its former fat self. At certain times of the month it disappears altogether. What is going on? Actually, the Moon always stays the same; it just appears to change shape.

As the Moon orbits Earth, light from the Sun strikes different areas of its surface; sometimes we can see a lot of its surface, sometimes only portions of it. When the entire surface of the side of the Moon that faces Earth is in sunlight, we see a *full moon*. However, when that side of the Moon is in shadow, we cannot see the Moon at all. We call this a *dark moon* or, more commonly, a *new moon*.

The Moon seems to be shape-shifting as it passes from full illumination to full shadow, and we call these changes its *phases.* The Moon has four phases. A few days after a new moon, the Moon *waxes* toward illumination again. If you look to the west about a half-hour after sunset, you will see a tiny sliver of moon called a *crescent moon.* The Moon appears to grow fuller as it slowly moves into the sunlight over the next week or so, until you can see about half of it. This is dubbed a *first-quarter moon.* Before long, all of the Moon is in sunlight and we see the *full moon.* Then the cycle reverses and the Moon begins to "disappear" again in a process called *waning.* It moves to a *third-quarter moon* and finally back to the full shadow of a new moon. This entire process takes a little more than twenty-nine days. One complete cycle of the Moon's phases is how we measure the length of a month.

MATERIALS

- notebook
- pencil

INSTRUCTIONS

Look at the weather page in your local newspaper to find out when the next new moon is. If you cannot find this information in the newspaper, check your local library. A couple of days after the new moon, go outside about a half-hour after sundown and look to the west. You should spot the sliver of a crescent moon. Sketch

everything in your line of view toward the crescent moon: the trees, houses, telephone poles—everything. Then add the crescent moon, being careful to get its position correct in relation to the other objects in your picture. Repeat this exercise every other day for two weeks. You don't have to redraw the house, trees, and other objects because they shouldn't have moved. Just draw in the position of the Moon and its shape. What do you notice? The Moon will have moved and changed into a full moon, and you will have charted its dance across the night sky.

If you want to chart the progress of the full moon turning

back into a new moon, you will have to get up really early to do so. You will have to go outside just after the Sun rises in the morning. Why? Because a waning moon is not even visible until late at night and is actually most easily viewed just over the western horizon very early in the morning.

SUN CLOCK

Warning! Never look directly into the Sun. Doing so can permanently damage your eyes. The only completely safe way to view the Sun is indirectly, by projecting its image onto a screen of some kind.

Introduction

Astronauts have traveled to the Moon, and they hope one day to travel to the planet Mars. In fact, it is the dream of many people (and a lot of science fiction writers) that before too long humans will permanently live in space in space colonies and from there eventually travel to the stars. But the star closest to Earth will never be a suitable destination for any space traveler because it is the Sun! That's right. Our Sun is a star, pretty much like all the other stars you see in the night sky. It appears so big to us because it is a lot closer to Earth than any other star—only 93 million miles away! And no matter how advanced our space technology, we will never visit this star because its surface temperature is 10,000 degrees Fahrenheit. That's hot! It explains why we can get a sunburn standing 93 million miles away.

If the Sun is so bright, why don't we see it at night? Good

question. The answer has to do with how Earth moves in relation to the Sun. The Sun can be thought of as stationary in relation to Earth. But Earth, like other planets, spins on its axis, like a top spinning on its tip. This motion is called *rotation*. Even though you cannot feel Earth rotating, it is spinning at about one thousand miles per hour. At that rate, it needs only about twenty-four hours to make one complete rotation on its axis. As Earth spins, the point on Earth where you are located faces the Sun and receives its light for half the turn (we call this daytime). Then the continuing rotation spins you out of the Sun's light for the other half of the turn (nighttime). To really understand this movement, try this demonstration. Imagine your body is Earth. Draw your country on a piece of paper and label the drawing with the north, south, east, and west directions. Then tape the drawing to your chest. Choose a lamp to serve as the Sun. Now face the Sun and begin to turn slowly to your left. Watch how the Sun travels across your country. As your right shoulder turns toward the lamp, the Sun will appear to set. As you turn your back to the Sun, your country will be entering night and you will not be able to see the Sun at all. As you continue to turn and your left shoulder once again turns to the Sun, day begins to dawn on your country. Makes perfect sense!

You can tell more than just whether it is day or night by noting where the Sun is. You can actually tell time according to the Sun's location in the sky. The Sun is low in the east in the morning. It is high overhead in the afternoon. And it is close to the western horizon near nightfall. But remember, looking at the Sun is very dangerous. You can permanently damage your eyes. You can tell just as much about the Sun's movement and location by looking at shadows, which are created by sunlight. In fact, you can use these shadows to make a clock!

MATERIALS

- piece of cardboard at least 8 by 10 inches
- spool
- 2 pencils (one to write with and one about 3 inches long for the shadow caster)
- watch
- glue
- a sunny day

Glue the spool to the middle of the cardboard. When the glue is dry, take the cardboard outside and put it in a sunny area. If it is windy, anchor the cardboard with small stones. Stick a pencil into the spool. Notice that the pencil casts a shadow on the cardboard—it looks just like the hand of a clock! Look at your watch and write the time on the cardboard at the tip of the shadow. As the Sun appears to move across the sky (remember, it's really Earth that is moving), the shadow will move. Each hour, go out to your Sun clock and record the time on the cardboard at the tip of the pencil's shadow. Pretty soon, you will have a Sun clock. To use this clock, all you need to do is look at where the shadow falls on the cardboard. Then read the time.

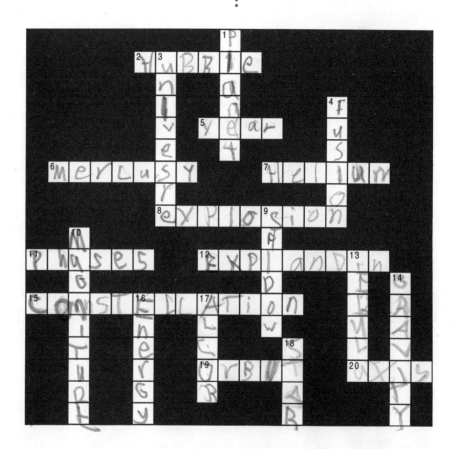

ACROSS

2 The highest powered telescope in the world— capable of viewing the dimmest stars.

5 It takes the Earth this long to travel once around the Sun.

6 The planet closest to the Sun.

7 When a nebula gets hot enough and dense enough, the hydrogen atoms inside it get squashed together to form this new element.

8 Many scientists believe that the universe began with an enormous one of these, called the big bang.

11 The Moon goes through _____ every 29 days.

12 Some scientists compare the universe to a balloon because, like a balloon that is being blown up, the universe is _____.

15 A group of stars creates a pattern that is known as a _____.

19 The planets do it around the Sun and the Moon does it around the Earth.

20 The Earth turns once on its _____ each day.

DOWN

1 Earth is one; so are Mars and Venus.

3 Although it may seem to be filled with stars, it is mostly empty space.

4 A nuclear reaction that takes place inside of stars.

9 A Sun clock uses a moving _____ to track the movement of the Sun.

10 The measure of a star's brightness.

13 A large cloud of gas and dust in which stars are formed.

14 The attractive force that pulls particles of gas and dust together to form stars.

16 A star releases lots of this in the form of light and heat.

17 Mizar's companion star. It is the dimmest star in the Big Dipper.

18 Our Sun is one of these; so is Alpha Centauri B.

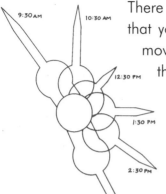

9:30 AM
10:30 AM
12:30 PM
1:30 PM
2:30 PM
3:30 PM

There is one not-so-small catch (besides the obvious fact that you cannot use this clock at night). You can never move this clock! If you move it, it won't work, because the shadows will not fall in the exact same places. This clock is not very convenient for your busy life, but you have to admit, it is cool!

OTHER THINGS TO TRY

Make a more permanent Sun clock using a big stick stuck into the ground. Mark the hours with stones.

F·U·N F·A·C·T·S

- If you drove a car to the Sun at a speed of 55 miles per hour, it would take you 193 years to get there. Hope you packed a lunch!
- Without the Sun, the temperature on Earth would be about 450 degrees below zero (Fahrenheit)!
- Our Sun is so huge that one million Earths would fit inside it.

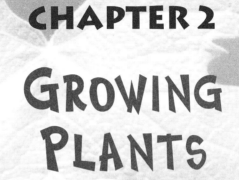

CHAPTER 2

GROWING
PLANTS

· ·

INTRODUCTION

Hey, don't look now, but there's a plant in your lap! It doesn't have leaves or flowers anymore, but it's there—in the fabric of your pants, skirt, or T-shirt. It's called cotton! And remember that pizza you ate the other day? Well, much of it comes from plants, too! The tomatoes, garlic, and wheat in the pizza are plant products (you can thank a cow for the cheese), and so is the cardboard box it came in. The wood in a baseball bat, the beans in a burrito, even the book you're holding this very minute—all are made from plants. Plants are everywhere in our lives. You couldn't get away from them if you tried. But why would you want to?

So, just what is a plant, anyway? Well, plants are living things. Unlike animals, however, they are "rooted" in place; they are stationary. Besides roots, most plants have stems, which provide the stability they need in order to grow up from the ground. They also have leaves, through which they absorb sunlight to make food. Most plants also have flowers and fruits, from which they produce seeds to make new plants.

We depend on plants for lots of things in our lives. For instance, we use the cellulose in plant

cells—which gives plants enough "stiffness" so they don't droop—to make cloth and paper. But mostly we rely on plants for food. We have plants to thank for the delicious fruits, vegetables, beans, nuts, cereals, and hamburgers we eat. Wait a minute— hamburgers? That's right. After all, hamburgers come from cows, and cows eat only plants. In fact, a cow has to eat sixteen pounds of plants for us to have one pound of hamburger! Nearly everything that provides natural food for people and other animals depends on plants. Just try to think of something that doesn't.

So, it's obvious enough that we eat plants. But what do plants eat? Mostly water and sunlight. Talk about a light meal! Because they can't move around, plants have to make their own food. They do this by trapping sunlight in their leaves. *Chlorophyll*—a pigment in plants that makes them green—is very good at absorbing sunlight, which the plant combines with carbon dioxide and water, and presto!—lunch is served. Actually, scientists have a much more dignified name for this process: *photosynthesis.* When plants make food, they are also helping us. You see, in order for plants to get carbon dioxide to make food, they have to inhale it from the air. Plants "breathe" through tiny holes on the underside of their leaves called *stomata*. Well, it's only logical that anything that breathes in must breathe out. And plants breathe out oxygen. Which is what we breathe in. Then we breathe out carbon dioxide, which plants breathe in. As you can see, plants "feed" us in more ways than one.

FLOWER PRESS

Roses are red, violets are blue. . . Okay, okay, we won't bore you with the rest of that rhyme. But stop and think about flowers for a moment. Take the rose, for instance. Roses aren't only red—they're yellow and pink and salmon and all kinds of other color variations. When you actually think about it, flowers are not only some of the most beautiful things in nature, they're also some of the most colorful. Ever wonder why? Well, a lot of the reason for the striking colors and patterns of flowers is that they serve a very important function in plants—they help plants reproduce. Basically, flowers are nothing more than seed factories. But they can't do their job if the pollen inside one flower never finds its way into a different flower. That's where beauty comes in—a flower's color and patterning can attract bees, butterflies, and other winged creatures.

If you look carefully at a flower, you will probably see a circle of several long, thin tubes protruding from its center. These are *stamens,* the male parts of a plant. The tips of the stamens are covered in powdery yellow *pollen,* which is necessary to fertilize another plant. When insects feed at the center of the flower, they get some of this pollen on their bodies. When they fly off to another flower, they deposit the pollen from the stamens of the first flower onto the *pistil* of the new flower. The pistil is the female part of a flower, and there is usually only one. It usually sticks up out of the very center of a flower. At the bottom of the pistil are several *ovules*, or plant eggs. When a pollen-covered insect lands on a flower, it may deposit some pollen on the pistil and the ovules. When the ovules are fertilized in this way, new seeds develop. The whole process is called *pollination*, and it all starts because roses are red and violets are blue. . . .

MATERIALS

- fresh flowers
- absorbent paper towels
- construction paper (optional)
- craft glue with water added
- a small board (such as cardboard or wood)
- several heavy books

INSTRUCTIONS Pick a bouquet of different kinds of flowers. The best kinds to press are ones that are not too thick. For example, roses are pretty thick, but daisies and buttercups are thinner and press very well. Before pressing the flowers, look at them carefully. Can you find the stamens, pollen, and pistil of each flower? Arrange a few flowers on an absorbent paper towel spread on a flat surface. Make sure the flowers do not touch each other. When they are all arranged, cover them with another piece of

pistil
pollen
stamen

absorbent paper towel. Then, place a board on top of the paper towel and weigh it down with several books.

After about two weeks, remove the books, board, and paper. Carefully lift the dried and pressed flowers from the bottom paper towel. If you place them on waxed paper that has been brushed with craft glue, then cover with one ply of a two-ply tissue, you get a beautiful translucent paper.

OTHER THINGS TO TRY

You can also press leaves, whole plants, and even seaweed, if you dry it off first.

FUN FACTS

To find out about rare and endangered flowers, write to your state's parks department or extension service.

Science.

SPROUTING SEEDS

Seeds are powerful little packets of life. Inside each seed is an *embryo*—the speck of life that will grow into a full-size plant. Did you know that nuts are actually seeds? For example, inside a walnut is all the information needed for the seed to grow into a walnut tree. But nothing can grow without food, not even a seed. So besides an embryo, a seed also contains a large food supply to nourish the embryo. It is because seeds are so full of "food" that you benefit from eating them. There's nothing like a handful of seeds, whether they be walnuts, sunflower seeds, or even popcorn kernels (after they're popped, of course!).

Many seeds can be a bit hard to chew until you crack open the shell. Actually, all seeds have some kind of hard outer layer, called the *seed coating*. The seed coating of a walnut is very thick, as is a coconut's. (Yes, a coconut is a seed!) The seed coating protects a seed from disease, from drying out, and sometimes from being eaten.

In some ways, a seed is like a chicken egg. Inside the egg is an embryo that can grow into a chick. The egg contains a food supply for the embryo—the yolk. And the shell protects the growing chick from the outside world, just

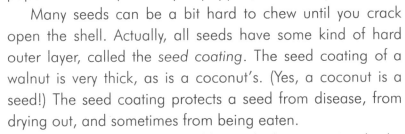

MATERIALS

- large (gallon-size) Ziplock bag
- paper towels
- dried seeds (dried beans, lentils, peas, birdseed, radish seed, plain popcorn, etc.)
- stapler
- tape

like the tough outer coating of a seed. Most of the chicken eggs that you buy at the store are not fertilized, so they would never grow into chickens. But a lot of the seeds you can get are fertilized, and you can very easily convince them to grow into a full-size plant. You can even grow seeds without dirt! That way, you can spy on every little step a seed goes through on its way to becoming a full-grown plant—from roots to shoots!

Lay two paper towels flat inside a resealable plastic bag. Try not to fold or crumple them. Next, staple across the bag in a row about four inches down from the top. The row of staples should go completely through the bag, and the staples should be spaced very close together, only $\frac{1}{8}$ inch apart. (You want only a small space between the staples through which the seeds' roots will grow.) Now, place the seeds inside the bag, above the row of staples. It is very tempting to put a lot of seeds in your garden, but if you overcrowd it, none of them will do well. A healthy garden will grow with ten to thirty seeds, depending on their size. For example,

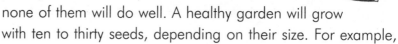

ten lima beans or thirty radish seeds are about right (or you can mix different kinds of seeds). Add ¼ cup of water and zip the bag closed. Tape the bag to a sunny window and wait. In about a week, you will see tiny roots beginning to emerge through the seed coatings. Soon after, you will see the stems poke out of the top of the seeds. At this point, you can transplant any healthy seedlings into soil and grow full-size plants. (Make sure the seedlings are large enough to transplant.)

OTHER THINGS TO TRY

Try to sprout many different types of seeds. For example, try cucumber, tomato, or orange seeds, and grow them in the plastic bag garden. Some seeds, like pine cone seeds, take a very long time to sprout.

Try watering the seeds with some liquid other than tap water. Make a guess about which liquids will help seeds grow and which will not.

Seeds need water in order to sprout. But is the Sun necessary? Do an experiment where you make two plastic bag gardens. Put one in a dark closet and one in a sunny window. Make a prediction about what you think will happen to each garden.

PLANT PIGMENT ART

If a plant were hungry, how would it eat? Would it (a) have a pizza delivered; (b) go to a fast-food restaurant; or (c) order out a sunbeam?

The correct choice is (c)—order out a sunbeam. Yes, plants eat in the wildest ways. They use the energy from the Sun, zap it around in the chlorophyll in their leaves, and turn it into food. It may not taste as good as pizza, but you can see the advantage: plants never have to go anywhere, and they can eat as much as they want whenever the Sun is out.

The chlorophyll helps the plant eat, but it is also a pigment, which is a chemical that makes a color. Chlorophyll gives the plant its distinctive green color. We have pigments, too, such as the pigment in our skin. The only difference between you and a plant is that you can't use the pigment in your body to make food from the sun. Oh, and you aren't green, either!

Not all plant pigments are green. Some are red, orange, yellow, and even blue. That is why some tree leaves look like they're turning different colors in the fall. The truth is, they always had all of those colors. It's just that the green pigment overpowers the red, orange, and yellow pigments.

In the fall, the green pigment breaks down and the other colors show through.

We love to look at the beautiful colors nature displays through plants. But we can also use those colors to make art.

(Adult Assistance Needed)

With an adult's assistance, cut the beet root into four parts. Measure two cups of water in a small pan. Add the beet root to the pan and simmer on low heat for half an hour. Remove the beet root and save it to eat later (it's good for you). The water in the pan will be full of beet pigment—a vibrant red.

Spread a square piece of cotton cloth flat on the table. Begin to twist the cloth as if you were trying to wring out imaginary water. Once you have it twisted, wrap rubber bands at various intervals down its length. The areas where the rubber band touches the cloth will remain white. If you want a mostly white tie-dye, put a lot of rubber bands on it. If you want a mostly red tie-dye, loosely twist the cloth and put only a few rubber bands on it. Place the twisted and banded cloth into the warm beet juice. Let it sit for 15 minutes,

and then remove it from the pan and let it dry. Be neat! Any beet pigment that spills could stain. No matter how tempted you are, don't unwrap the cloth until it is dry. It may take all night, but it will be worth the wait.

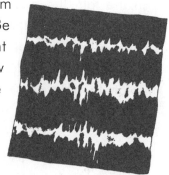

Try making plant dyes with other plant parts, such as onion skins, dandelion roots, goldenrod flowers, or blackberries.

Make a multicolor tie-dye by twisting up an already dyed (and dried) piece of cloth, securing it with rubber bands, and dying it in a different color plant pigment.

While some plant pigments such as chlorophyll help the plant make food, other pigments have other jobs. Some function as a sun block to protect the plant's food-making pigments from getting burned by the Sun, and others work as a chemical "perfume" to attract helpful insects to the plant (for pollination, for instance).

SODA BOTTLE BOTANY

Do you know someone who loves houseplants but always forgets to water them? With help from plant biology and an old soda bottle, you can solve this problem. In fact, you can build a plant chamber that will hold a plant for several months without ever needing additional water! In this kind of closed system, plants can keep using the same water over and over again. In effect, the plants will water themselves!

No matter how hard you search, you won't find a plant's mouth. That's because it doesn't have one. Instead, plants drink through their roots. The roots reach deep down into the soil and spread out in search of water. When water comes in contact with the roots, it gets pulled into the plant through root hairs. Root hairs look just like—you guessed it—hairs! Root hairs are extremely short and thin, so water can pass through them and into the plant very easily.

Once inside the plant, the water travels through it in little pipes called *xylem*. The xylem are also very small and are hard to see. (But try this: Cut a stalk of celery in half and look carefully at the cut end; you may be able to see a bundle of xylem.) Eventually, any water the plant doesn't use exits it through tiny holes on the undersides of the leaves. (The holes are called *stomata,* and the plant also uses them for breathing.) The water leaves in the form of water vapor, which is like nearly invisible clouds. As the water vapor exits the leaves, suction pulls more water up into the roots, so the whole plant acts rather like a straw! This entire process is called *transpiration.*

In the out-of-doors, the water vapor from plants drifts upward to the clouds. When it rains, water goes right back down into the soil to be used once again by the plant roots. In your soda bottle, visible clouds won't form. But you will see water vapor droplets forming inside the top of the bottle. This is where the little invisible clouds turn back into visible water. The water droplets cling to the sides and top of the bottle, and eventually slide down into the soil and water the plant. As this process repeats itself over and over, the plant stays nicely watered.

MATERIALS

- clear plastic soda bottle (1-, 2-, or 3-liter) with a reinforced bottom
- scissors
- hot water
- soil
- tape
- any small plant (spider plant, African violet, ivy, coleus—just about any plant will do)

INSTRUCTIONS

(Adult Assistance Needed)

Soak the soda bottle in hot tap water. The hot water will loosen the label and the reinforced bottom. Remove the labels and the bottom. (Sometimes you have to use a little muscle to

pull the bottom off.) Once you get the base off, tape up the little holes in its bottom and fill it with moist (but not really wet) soil. Plant your plant in the soil, pushing the soil down firmly so there will be no large air holes around the roots.

With an adult's help, carefully cut off the top $\frac{1}{4}$ of the soda bottle near where the bottle begins to get narrow. Turn the bottle upside down over the plant in the base, fitting the bottle inside the plastic base. Then, tape them securely together. Now your plant is sealed in a closed environment. Set the plant in a sunny window and begin your observations. How long does it take for water vapor droplets to appear on the sides of the bottle? What time of day do you most often see these water drops? Are there more water drops on sunny days or rainy days? Does the weather even matter? Does temperature matter? Keep a journal of your observations and see if you detect a pattern to the transpiration cycle.

The crossword puzzle solution:

1 Across: OVULES
2 Down: LEAVES
3 Across: CELLULOSE
4 Down: EMBRYO
5 Down: SEEDS
6 Across: STAMENS
7 Down: STEM
8 Across: ROOTS
9 Down: PIGMENT
10 Across: PHOTOSYNTHESIS
11 Down: OXYGEN
12 Down: SUNLIGHT
13 Down: TRANSPIRATION
14 Down: EGG
15 Down: XYLEM
16 Across: POLLEN
17 Across: CHLOROPHYLL

ACROSS

1 "Plant eggs" located at the bottom of the pistil.
3 The part of a plant cell that gives a plant its stiffness. We use this substance to make cloth and paper.
6 The male parts of a flower.
8 The part of a plant that grows underground and takes in water and nutrients from the soil.
10 The process by which plants make their own food.
16 A powdery yellow substance that covers the tips of the stamens.
17 The substance in plants that absorbs sunlight.

DOWN

2 Plants absorb sunlight through their _____.
4 The speck of life inside a seed that can grow into a plant.

5 When an ovule is pollinated, it becomes one of these "packets of life."
7 A plant's _____ provides stability.
9 Chlorophyll is a _____ because it makes plants green.
10 The process by which an ovule is fertilized with pollen.
11 A waste product of photosynthesis that animals need to survive.
12 Plants need water and _____ to make food.
13 The process by which water enters a plant through the roots and travels upward to the leaves, where it exits through tiny holes.
14 A seed is a lot like an _____.
15 The pipes inside a plant through which water travels.

CHAPTER 3

UNUSUAL MIXTURES

● ●

CRYSTALS
....
POLYMERS
....
ACIDS, BASES, AND PH
....
MAKING SOAP, NO LYE!
....

INTRODUCTION

Everything is made out of matter—from a star at the farthest reaches of the universe to the shoes on your feet. Everything! Asteroids, rain, book reports, puppies, eggs, oxygen, video games, teddy bears. You name it—it's matter! *Matter* is anything that takes up space and has a measurable amount of "stuff" (called *mass*). That means you are made of matter, because you take up space and you can be measured in pounds. Chemistry is the study of matter.

Most people don't think in terms of matter. If you ask someone what a teddy bear is made of, he or she will probably tell you it is cloth and stuffing. But you can tell them that a teddy bear is made of matter, because cloth and stuffing take up space, and the amount of stuff they are made of can be measured.

Let's take a closer look at the teddy bear and break it down into its tiniest parts. For example, the cloth and stuffing are sometimes made from cotton, which is a plant. So you could say that the teddy bear is actually made of plant parts. Okay, so what are plant parts made from? They are made from plant cells. Plant cells are the smallest part of a plant, and they are so small you can only see them with a microscope. But guess what? Cells are made of something even smaller—tiny things called *molecules.*

Molecules are so incredibly small that you can't see one just by looking at it with your eyes. You have to use all kinds of fancy scientific equipment to see a molecule, because it is a lot smaller than the point of a pin! You might think that a molecule is the smallest possible bit of matter, but it isn't. Molecules, it turns out, are made of *atoms.* And atoms are almost the smallest things that scientists know about. But even though atoms are extremely small, they are still made of matter, because they take up space (just a very small amount of space) and their amount of "stuff" can be measured. Atoms, as you may have guessed, are made of even tinier parts, called *quarks.* Scientists are just learning about quarks now. Do you think quarks are made of something even smaller? No one knows for certain.

There are about one hundred different known atoms. When similar atoms bind together, they form *elements*—such as gold, helium, aluminum, and oxygen. When different kinds of atoms bind together, they form molecules—such as water, salt, and air molecules. There are nearly countless ways to combine atoms. Just think about all the different forms of matter that exist in our universe!

A CHANGE IN MATTER

We said that an egg is matter. But what happens to the molecules in an egg when it is cooked? Are the molecules of a fried egg still egg molecules? You bet! Although the physical appearance of the egg changes, the molecules do not change. What happens is that the heat rearranges the relationships between the molecules. The egg goes from liquid to solid, but it's still an egg! In chemistry, this is called a *physical change.*

Chemical changes are different from physical changes. A *chemical change* occurs when molecules change into completely different molecules. For example, try adding one teaspoon of vinegar to one teaspoon of baking soda. The molecules in the vinegar react with the molecules in the baking soda, and together they make new and different molecules. When the molecules of the vinegar and those of the baking soda break apart, some of them recombine to make a very familiar molecule: water!

You can try this experiment because it is completely safe. You won't be able to see the new molecules (they are too small), but you will definitely see some exciting chemical changes!

CRYSTALS

Imagine how rich you would have to be to brush your teeth with diamonds! Okay, maybe that isn't such a good idea. But diamonds and toothpaste do have something in common. They both contain crystals. The world is filled with interesting crystals. The salt for your french fries and the sugar for your cereal are crystals. The "sand" on sandpaper is made from the same crystals that make rubies and sapphires. Without crystals, we wouldn't have snow, vanilla flavoring, computers, video games, radios, or diamond rings.

Let's make it crystal clear just what a crystal is. It is a mineral whose atoms are arranged in very orderly geometric patterns that are repeated over and over again. At the atomic level, a crystal is as orderly and symmetrical as a snowflake! Although these often intricate patterns are not visible on the outside of a crystal, they are often suggested by the beautiful symmetry of the "faces" of a crystal. *Symmetry* means that all the sides have the same shape and are opposite each other.

Despite their intricate geometrical patterns and unique

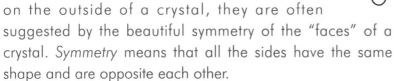

MATERIALS

- 1 cup of granulated sugar
- ½ cup of water
- small pan
- wooden spoon
- small glass jar (suitable for canning)
- pencil
- string
- paper clip
- food coloring (optional)

properties, crystals are not very hard to grow. Let's give it a try. In fact, let's make rock candy crystals. That way, you can grow your crystal and eat it, too!

(Adult Assistance Needed)

Mix the sugar and water together in the pan. Stir with a wooden spoon until you don't think that any more sugar will dissolve in the water. Then, ask an adult to help you heat the sugar and water over low to medium heat on the stove. Keep stirring, and in less than five minutes, all of the sugar will be dissolved in the water. (All the sugar crystals are now tiny individual molecules that you can't see). The mixture should look almost like water, only a little thicker. Turn the stove off, remove the pan from the stove, and let it cool for about five minutes. At this point, add five or six drops of food coloring so the rock candy crystals will be beautifully colored when they grow.

While the mixture is cooling, set up your rock candy crystal growing chamber. Cut a piece of string about ten inches long. Tie one end to the pencil and the other end to a paper clip. Put the string in the

jar so that the paper clip touches the bottom and
the pencil sits across the mouth of the jar. Wind
the string around the pencil so that it doesn't
droop inside the jar. When the sugar and water
mixture has cooled a little, pour it carefully into
the jar. Now let nature take its course. It will take
a few weeks for the water to evaporate and the
crystals to form. Try not to disturb the jar while
the crystals are forming. Once they're grown, examine them
for their unique symmetry.

Try the same recipe, but replace the sugar
with Epsom salt. How are these crystals
different from the rock candy crystals?
Don't eat the Epsom salt crystals!

F·U·N F·A·C·T·S

Ancient Greeks used the word *krystallos* to mean both ice and quartz. They thought that quartz (which is a very common crystal) was a kind of permanently frozen ice. Amethyst is a purple-colored quartz crystal. In Greek, "amethyst" means "not drunk." The myth is that whoever wears amethyst jewelry is protected from the effect of too much alcohol. Do you believe it?

Some kinds of crystals, such as quartz, have very intriguing physical properties. When specially cut quartz is zapped by small bits of electricity, it bends. When electricity zaps the quartz over and over, the quartz vibrates. The vibration is so consistent that it is used to control clocks and watches.

POLYMERS

Take a look at yourself right now. Are you wearing any plastic? Maybe you can see some plastic on the tips of your shoelaces. How about the buttons on your shirt? Look some more. You're probably wearing lots of plastic and don't even know it! For instance, did you know that most sweatpants and sweatshirts have some plastic in them? So do many T-shirts. Some fleece jackets are made out of recycled plastic soda bottles!

All of the plastics you are wearing are *synthetic polymers*. These are substances made by humans. There is another type of polymer that is made in nature. Not surprisingly, this is called a *natural polymer*. Your fingernails are not made out of plastic, but they are also polymers—natural polymers. Some other natural polymers are rubber, egg whites, wool, porcupine quills, and lobster shells.

Rubber and porcupine quills seem so different that you may be asking yourself, "What on earth is a polymer?" We thought you'd never ask! A polymer is made of many molecules strung together like a chain of paper clips. Pretend that each paper clip is one molecule, and that each molecule is exactly the same kind of molecule as the one next to it. When you clip a lot of the paper clip "molecules" together, you make a model of a polymer. Sometimes polymers are long skinny strings like a paper clip chain. But often the long chains loop around and are clipped together along the middle, rather like a ladder. Still other polymers are long chains with lots of side chains branching off in every direction.

No matter what shape they take, individual polymers are very tiny. They have to be bound together in a big tangle in order to see them.

You can make your very own polymer called polymer putty. This putty does lots of cool things—it bounces and stretches in weird ways. It is made by means of a chemical reaction. Remember what that is? It's when the molecules of two or more different substances break down and recombine to form an entirely different substance from the two original substances. Here's how it works. Elmer's glue is made of a polymer. When it is mixed with water, the chains of polymer molecules spread out. When you add borax, a chemical reaction takes place. The molecules in the borax hook the chains of molecules in the glue together at various points to form a ladder shape. This shape is much less flexible than the chains, so the new polymer forms as a putty instead of as a liquid glue.

MATERIALS

- 1 tablespoon Elmer's Glue™
- 1 teaspoon borax
- water
- 2 small cups
- measuring spoon

INSTRUCTIONS

In one cup, mix one tablespoon of Elmer's glue with one tablespoon of water and stir well. Put the cup aside for a moment. In the other cup, mix eight tablespoons of hot tap water with one teaspoon of borax and stir until the borax is dissolved. Now, measure one teaspoon of the borax and water mixture and stir it into the cup containing the glue and water mixture. Stir quickly! Chemistry is amazing—right before your eyes you will see the polymer putty appear!

Keep stirring. After about one minute of stirring, put the polymer putty on a plate and let it dry for a minute or two.

Now you can start experimenting with the polymer putty. Very slowly, try to pull it apart. How far will it stretch? Roll it into a ball and bounce it on a hard surface. How high will it bounce? Flatten the putty and press it across a piece of newspaper. When you peel it back, what do you see? Experiment with the putty to discover all the crazy ways this substance behaves.

NOTE: Polymer putty is not hazardous with normal use. However, do not eat it or use it near your face. It can stain, so watch where you put it. Wash your hands after handling it. After making the putty, any liquid left in the cups can be rinsed down the drain. If stored for a long time, polymer putty can develop mold. To prevent this, store it in a plastic bag in the refrigerator. If mold grows on your polymer putty, throw it out and make a new batch.

ACIDS, BASES, AND PH

Acids

Some people drink a glass full of acid at breakfast every morning. Do you? If you drink orange juice, then you're drinking acid. It's not very strong acid, not like the vats of acid that bad guys always fall into in the movies. No, orange juice is a weak acid. And it's not the only acid that you regularly come into contact with. There's also acid in your stomach and muscles, and acid is even in aspirin and cheese.

How can something belong to the acid club? All acids, whether they are weak, like orange juice, or strong, like sulfuric acid, contain *ions*. Ions are tiny particles that have either a positive charge or a negative charge, like the ends of a battery. Acids contain a specific type of ion called *hydrogen ions,* which are positively charged.

Bases

Just as big has small and up has down, acids have *bases.* Bases are the opposite of acids. They contain negatively

ACTIVITY

charged particles called *hydroxyl ions*. Some famous bases that you might know are baking soda, ammonia, and soap! Just like acids, bases can be weak or strong. One kind of strong base is called *sodium hydroxide,* also known as lye, which would burn a hole in your hand if you let it.

A Helpful Way to Look at Acids and Bases

Some people might say, "So what! Acids have little positively charged hydrogen ions floating around in them, and bases have little negatively charged hydroxyl ions. Who cares!" Well, acids and bases may seem a little dull on their own, but an exciting way to think about them is to think about how they react with things. The acid in acid rain destroys trees, lakes, and even statues! The acid in your stomach helps to break apart the food you eat, so your body can use it. The base in baking soda makes the cake batter rise. The base in soap helps clean dirty hands and dishes. It is safe to say that acids and bases are very active substances.

Mixing Acids and Bases Together

You don't have to be a mad scientist to wonder what would happen if you mixed an acid and a base together. Well, if the acid and the base were of equal strength, they would balance each other. The positively charged hydrogen ions and the negatively charged hydroxyl ions would pair up, one by one, like partners in a square dance, and the solution would become neutralized. A neutral substance has no acids or bases, and it can't dissolve statues or make cake batters rise.

Testing Acids and Bases

Safety warning: Do not ever taste or touch something if you don't know what it is!

So how do you know if something is an acid or a base? Some weak acids, like orange juice and lemonade, taste sour, but most acids are very dangerous, so we cannot use touch to know whether something is an acid. Some weak bases, like soap, feel slippery to the touch. However, most bases are very dangerous and can seriously burn you, so we cannot use touch to tell whether something is a base. Fortunately, there are some almost magical potions called *indicator solutions* that scientists use to test acids and bases safely. You can test some of the substances around your house with an indicator solution, which is really easy to make from—believe it or not—a purple cabbage! When this indicator solution is added to an acid, a chemical reaction takes place and the liquid turns red. When it is added to a base, it turns green.

Acids and bases can be compared on a pH scale, which goes from 0 to 14. Strong acids have very low pH numbers, whereas strong bases have high numbers. Substances like plain tap water have a pH of 7, meaning that they are neutral.

MATERIALS

- ¼ of a small purple cabbage
- 3 cups of water
- pan
- strainer
- several glass jars
- vinegar
- baking soda

I-N-S-T-R-U-C-T-I-O-N-S

(Adult Assistance Needed)

Have an adult help you chop the cabbage, and put it in the pot with the water. Simmer on low heat for about ten minutes. Take the pan off the heat and let it cool for five minutes. Strain out the cabbage juice and save it in a jar.

Now, here is the fun part. Put one tablespoon of vinegar in another glass jar. In a different glass jar, mix one teaspoon of baking soda in one tablespoon of water. Spoon two tablespoons of purple cabbage juice into

F-U-N F-A-C-T-s

- A bee sting is acidic and can be neutralized with a little baking soda. However, a wasp sting is basic and can be neutralized with a little bit of vinegar.

each of these mixtures. The indicator turns pink or red when it detects an acid, and it turns blue or green when it detects a base. Read the pH scale to see the pH readings of many common substances. Also, try testing other items such as dishwashing powder, lemon juice, cola, cream of tartar, or Alka Seltzer.

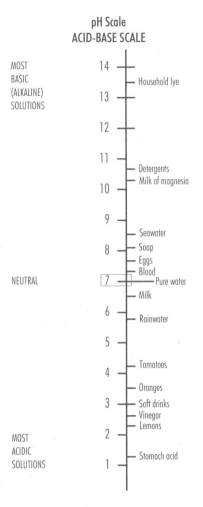

pH Scale
ACID-BASE SCALE

MOST BASIC (ALKALINE) SOLUTIONS

14
— Household lye
13
12
11
— Detergents
— Milk of magnesia
10
9
— Seawater
8 — Soap
— Eggs
— Blood

NEUTRAL
7 — Pure water
— Milk
6
— Rainwater
5
— Tomatoes
4
— Oranges
3 — Soft drinks
— Vinegar
— Lemons
2

MOST ACIDIC SOLUTIONS
1 — Stomach acid

MAKING SOAP, NO LYE!

Imagine living in a world without soap! No one would hound you to wash your hands before dinner or to do the dishes after dinner. Sounds pretty good, doesn't it? But wait a minute. Imagine how you would smell after a month without a soapy bath. What about all the germs that would accumulate on the dishes, forks, and spoons? Eating with your hands wouldn't be a help if you hadn't washed them in months. Maybe a world without soap isn't such a good idea after all.

Ancient peoples knew how to make a crude type of soap, but it was difficult to make and was not very widely used. As a result, the ancient world was not a very healthy or nice-smelling place. Without soap, people often got sick because they couldn't keep themselves or their environment clean. It wasn't until about three hundred years ago that people figured out how to make soap easily and in large quantities. First, they burned logs in order to get ash. Then, they mixed the ash with animal fat over a fire. As the ash and fat mixture got hot, it underwent a chemical reaction. This means that the molecules in the ash and fat changed into an entirely new substance—soap.

However, once you have soap, you can keep making it without having to use a chemical process. Some soaps are simply mixtures—that is, they are a lot of different substances mixed together. For example, when you bake oatmeal cookies, you mix flour, sugar, oats, and raisins. As the mixture bakes, all these ingredients melt together. But they don't undergo a chemical reaction and change into anything different. The raisin stays a raisin! And if you analyzed the cookie really closely, the sugar, oats, and flour would all be identifiable in the mixture. You can make soap this way, too. All you have to do is mix together finely ground soap, water, glycerin, and alcohol. Let's try it.

MATERIALS

- ¼ of a personal-size bar of Ivory Soap—(about ⅓ cup, grated)
- 1 tablespoon vodka (have an adult help you with this)
- 1 ½ tablespoons glycerin (can be bought at most drugstores)
- 1 tablespoon water
- cheese grater
- small pan (2–5 cup capacity)
- spoon
- containers to use as soap molds (such as empty yogurt containers, muffin tins, candle molds, etc.)
- stove
- fragrances (such as perfumes or colognes)
- food coloring (optional)

(Adult Assistance Needed)

Carefully hold the soap against the side of the cheese grater with the smallest holes, and grate it. (Be careful not to grate your fingers! You may want an adult to do this part for you.) The soap should be grated into a fine powder. Put the grated soap in the small pan. Add the water, vodka, and glycerin, and stir everything together. (Be careful when adding the vodka

because it is flammable.) Put the pan on the stove and simmer the mixture over very low heat. It will look clumpy at first. Don't worry. Stir the mixture occasionally for the first five minutes. Then, stop stirring and let the mixture simmer slowly for another five minutes. Take the pan off the stove and stir in a few drops of fragrance and food coloring. Rub the mold with a little glycerin so the soap won't stick. Then, carefully pour the soap mixture into the mold. Set the mold aside for half an hour until it hardens. To get the soap out of the mold, stick a butter knife between the soap and the mold and loosen around the edges. Then, carefully lift the soap out. Now, go wash your hands!

OTHER THINGS TO TRY

Try adding other ingredients to the soap mixture. For example, add vitamin E oil, aloe, or honey. Then, you can compare the soap bars to determine which cleans the best and which you like the best.

FUN FACTS

When you wash with soap that contains glycerin, some of the glycerin remains on your skin after washing. Glycerin attracts moisture, so your hands won't dry out. Substances that attract moisture are called *humectants*.

ACROSS

6 The first soap was made from a mixture of animal fat and _____.

8 A substance that attracts moisture.

11 A mineral whose atoms are arranged in an orderly, repetitive, geometric pattern.

14 When atoms that are the same bind together they form an _____.

15 Atoms are made of these tiny particles.

18 When purple cabbage juice is used to determine if a substance is an acid or a base, it is called an _____.

19 When a substance changes from a liquid to a solid, it undergoes a _____ change.

20 When all of the sides of an object have the same shape and are opposite each other it is described as _____.

DOWN

1 A substance containing negatively charged hydroxyl ions.

2 Anything that takes up space and has mass.

3 When substances react with each other to form completely different molecules, it is called a _____ reaction.

4 The ancient Greeks believed that quartz was a form of this very cold substance.

5 A measurable amount of stuff is called

_____.

7 When different types of atoms bind together they form these.

9 When ingredients are combined without undergoing a chemical change, the resulting substance is a _____.

10 Plastic is a _____.

12 The ancient Greeks believed this purple crystal prevented drunkenness.

13 The smallest units of matter that can come together to form elements.

16 Vinegar, which contains positively charged hydrogen ions, is an _____.

17 When hydrogen and hydroxyl ions join together, they become this common substance.

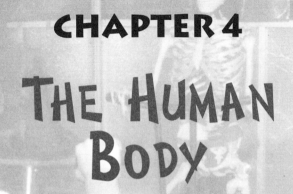

CHAPTER 4

THE HUMAN BODY

● ●

INTRODUCTION

What runs on fuel, can accelerate around corners, stops on a dime, and comes in many colors? We're not thinking of a race car. Here's another clue. It has an intricate electrical system powered by a brain. Give up? It's you! The human body is one of the most complicated "machines" in the world. Like a fancy sports car, in order to keep "running," all the body's complicated systems have to run smoothly and in sync. Let's take a closer look "under the hood."

FUEL

In order to run, a car has to convert gasoline or some other fuel into energy for power. The human body uses food as its fuel. The average adult eats about three pounds of food a day. How much do you eat?

The fuel in a car has to be directed to the engine if the car is to be able to go anywhere. To get there, the gasoline is pushed through tubes by a fuel pump. The nutrients from the food you eat are absorbed into your blood, and this fuel is brought to every part of your body as your blood circulates. The fuel pump in your body is your heart, and the fuel is in your blood. The tubes your heart pushes the blood through are your blood vessels. The blood also carries many other important things: for example, oxygen, waste products, and germ-killing cells.

ELECTRICAL SYSTEM

Cars today are full of electric components such as headlights, power windows, and radios. But do you know that you have an electrical system, too? It's called your *nervous system.* Electrical impulses are sent over the nerves in your nervous system to tell your body what to do to stay alive! Things like breathe, eat, and move.

Your brain is the electrical control center that rules your nervous system, controlling everything your body does. Your brain enables you to think, see, smell, hear, read, laugh and cry, and a zillion other things. This amazing organ weighs only three pounds and is mostly made of water (85 percent), but it is responsible for all the complicated and creative things you think and do.

CHASSIS OR FRAME

Car makers want to make safe cars, so they work

hard to make long-lasting, strong frames for their cars. But no matter how hard they try, they will never invent a car frame that matches the strength and durability of your frame—your skeleton. Your frame is made of bone, which is living material. When a bone breaks, it can usually repair itself. Even over the course of a lifetime, most bones don't wear out. By comparison, a car frame may be ready for the scrap heap by the time it reaches the age of a teenager!

We don't often think about our bones, but just imagine your body without them. It's pretty hard to do, because your body would have no shape. Bones give your body structure and provide protection. Without the bones that protect your brain, you could give yourself a massive headache just by putting on a hat!

Paint Job

Like cars, people come in many different and beautiful colors. In animals (humans are animals!), skin color is caused by pigment. If you have darker skin, you have more pigment than someone who is light-skinned. The pigment in your skin can protect you from the dangerous *ultraviolet* rays of the Sun. When skin is exposed to sunlight, more pigment is produced, making the skin darker and protecting it somewhat from further ultraviolet damage. Of course, producing pigment is a slow process, and many people with lighter complexions try to hurry the process by getting their dose of sunshine in a short time. These people end up with a painful and unhealthy sunburn. In general, people who live in very sunny areas, such as tropical America, Australia, and Africa, have a lot of pigment, which makes their skin darker and protects them most of the time.

FUN FACTS

Some reptiles and marine animals can change the pigment in their skin so quickly that they can change from green to red or blue in seconds!

STETHOSCOPE

Ask your friends to name the strongest muscle in their body, and some may tell you their arm muscle is strongest. Others may choose their powerful leg muscles. But they'd all be wrong. Although arm and leg muscles really are strong, they are weaklings compared with your heart. That's right—your heart! Your heart is a muscle that started flexing before you were born and that won't stop until you die. To get some idea of what a remarkable feat that is, try flexing your leg muscles for ten minutes by doing deep knee bends. Oooowww! Now, you can really appreciate the job your heart does.

What's more amazing is that your heart is practically hollow! Well, actually, it's filled with blood. When your heart flexes, it squeezes blood out into the rest of your body. As the blood leaves the heart, new blood flows in. The blood is like a conveyor belt that carries oxygen and nutrients. It makes deliveries to all the cells in your body that need these things. This conveyor belt also makes pickups. When a body cell has waste it needs to get rid of, the blood will make the pickup and haul these wastes away.

ACTIVITY

After making deliveries, the blood returns to the heart for a quick pit stop. Then it goes over to the lungs, where it picks up oxygen. Every part of your body needs oxygen, so the blood has a full load when it heads back to the heart to be directed on its next run through the body. Nutrients are usually carried in the *plasma*, while oxygen is carried by the red blood cells. Your blood contains other kinds of cells as well.

Your heart is a workaholic—it keeps the blood flowing while you're awake and while you're asleep. Every day, all day long, it pumps blood to every nook and cranny of your body. Over and over, the cycle is repeated. In fact, all of the blood in your body gets pumped through your heart about one thousand times every day. Not a bad day's work for an organ the size of your fist that weighs about half a pound!

Even though you and your heart are inseparable, you probably pay little attention to it. After vigorous exercise, we try to catch our breath and say we can "hear" our heart pounding. But we're actually feeling it more than we're hearing it. However, there is an easy way to hear your heart. Doctors do it all the time—they use a stethoscope.

MATERIALS

- rubber tubing or an old hose that fits over a funnel or mouth of a plastic soda bottle
- a funnel or the top half of a plastic bottle

Tape the tube to the neck end of the funnel or plastic bottle. Place the end of the tube gently against your ear (do not stick it all the way into your ear), and put the flat part of the funnel or plastic bottle against your chest. Listen carefully. Can you hear your heart pumping away? Take a deep breath; you should be able to hear the air flowing into your lungs. Place the stethoscope against your throat as you swallow. Can you hear the saliva slide down your throat? Place the stethoscope on your head. Can you hear yourself think?

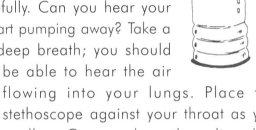

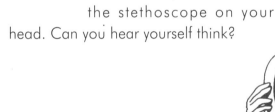

OTHER THINGS TO TRY

Try listening to other people's hearts. Or listen to a cooperative dog's or cat's heart. Do their heartbeats sound different from your heartbeat?

Listen to your heart while you are sitting. Put the stethoscope down and do twenty jumping jacks, then listen to your heart again. What difference do you hear?

What other things might work as a stethoscope? How about a paper-towel tube? A paper cup? What else?

F·U·N F·A·C·T·S

You have about 60,000 miles of blood vessels in your body.

MUSEUM OF Science.

LUNGS

Did you know that there are two wet, spongy, rather slimy things moving in your chest? They're your lungs. They may not be pretty, but you wouldn't be alive without them.

Lungs, as most of us know, are for breathing. They extract the oxygen from the air we breath and make it available for use by the rest of our body. No oxygen, no life. So, if you want to stay alive, breathing is very important. But have you ever stopped to think about how air gets into your lungs?

Just because you have a clear nose or an open mouth doesn't mean fresh air is going to fly down your trachea (your air tube) and into your lungs. You have to pull the air in somehow. To explain how you do this, we have to take a tour of your lungs. Your lungs are in a part of your chest that is a little like a small room with a rubber floor. This "room" is called the *thoracic cavity*. Your two lungs (which are about the size of two footballs) live in this room. It has two doors— your mouth and your nose. The rubbery floor—technically called the diaphragm— is made of muscles. The neat thing about this room is that it isn't one fixed size. It gets bigger and smaller as the "floor" muscles expand and contract. As

ACTIVITY

MATERIALS

- 1 small, plastic soda bottle
- 2 balloons
- 2 rubber bands
- scissors

you breathe in, the floor of the room lowers, making the room bigger, causing air to be pulled into your nose or mouth. As the chest muscles contract when you breathe out, the floor rises, making the room smaller and forcing air out the doors of your mouth and nose. That's breathing!

You can't see your lungs in action, but you can feel them by placing your hand on your chest as you breathe. You can also build a model of your lungs that is pretty cool. Try it, and you'll see that we're not full of hot air!

INSTRUCTIONS

(Adult Assistance Needed)
Cut the bottom from a small, plastic soda bottle. Stuff a balloon down through the mouth and into the neck of the bottle. Secure the balloon to the bottle by slipping the lip of the balloon over the mouth of the bottle. If it is loose, secure it in place with a rubber band. This balloon represents a lung. (Most people have two lungs, but for our model we'll need only one.) Next, cut the neck off the other

balloon. Stretch the cut balloon over the open end of the soda bottle like a drum skin, but not as tight. Secure the balloon in place with a rubber band. This balloon represents the "rubber floor" of the thoracic cavity. Now gently pull the diaphragm out toward you. You will see the lung "inhale." Then push on the diaphragm. Watch the lung "exhale." That's how your lung works!

FUN FACTS

If you've ever threatened to hold your breath until you die if you don't get that new bike—give up! You can't hold your breath until you die. Your body won't let you. Your brain knows what's best for your body, so it takes control. You faint and start breathing. Please take our word for it and do not try this!

YOUR SENSE OF SMELL

Have you ever smelled brownies just as they came out of the oven? They smell so good that you don't even wait for them to cool before you eat one or two! Mmmmm! The gooey chocolate tastes scrumptious as it melts in your mouth. Now, imagine a different scene. You are in bed with a bad cold. Your nose is stuffed up and you feel pretty miserable. Your mother, who is worried that you aren't getting the proper nutrients because you don't feel like eating, brings you a bowl of chicken soup and a brownie for dessert. How do you think that brownie will taste now?

Probably not very good. The fact is, it isn't your tongue that does most of the tasting; it's your nose! So when your nose is plugged, food just doesn't have the same appeal.

Noses are made for smelling. They do their job very well, and they help out the tongue when it comes to tasting. That's because a lot of what we call taste is actually a smell. Simply put, food tastes so good because it smells so good.

The reason your nose helps you taste is that the *olfactory nerves* found high in your nose are directly linked to your brain. When the olfactory nerves detect a smell, they signal your brain to decode it. In addition, in the back of your mouth there is an opening that goes straight up to the olfactory nerves. So when you eat something, the smell of the food moves directly to the olfactory nerves, which send the smell signal to your brain, which decodes it: Yahoo! A double-fudge brownie!

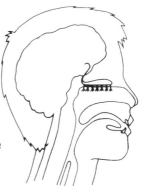

Place a handful of different-flavored gumdrops on a dish. Find a partner and have your partner close his eyes and pinch his nose closed. Now select a flavor of gumdrop and have your partner eat it. Make sure his nose is pinched close. Will he be able to tell you what flavor gumdrop you just served him? Chances are, he won't. It's nearly impossible to sense flavor when you can't smell it. However, he'll be surprised to see how well he can taste the gumdrop as soon as he unplugs his nose. Switch places now, and you try the experiment.

Instead of gumdrops, use pieces of fruit and vegetables. A raw potato and an apple feel the same in your mouth (they have the same texture), but do they taste different when you eat them with a pinched nose?

Try chewing a piece of onion with your nose pinched!

Did you know that flies taste with their feet?

THAUMATROPE

You have a twinkle in your eye. He gave me the evil eye. I can see it in your eyes. Irish eyes are smiling! The eyes are the windows of the soul. We have a lot of sayings about eyes because our eyes are very expressive. But did you know that you don't see only with your eyes? You also see with your brain. The eyes may be your windows on the world, but the brain is the computer that interprets what you see.

Think of your eyes as collecting instruments—they collect information. To be a little more specific, the eyes focus the light rays from the objects that we "see" in the world. The only things that your eyes sense are these light rays. They can't actually tell the difference between a train barreling down the tracks right at you or a puppy running toward you with a rubber ball in its mouth! It's your brain that tells you what those patterns of light rays are that the eyes are sensing. The signals that our eyes send to the brain are interpreted almost instantly as objects that we see.

Sometimes our eyes play tricks on us. We really should say our brain plays tricks on us. For instance, an optical illusion is a visual trick that happens because our brain can interpret what we're seeing in more than one way. This happens all the time in our lives. For example, every time you watch television, your brain is "filling in" information that your eye can't possibly be seeing. The picture on a TV screen is nothing but tiny dots vibrating very fast. But your brain "sees" people and backgrounds—a sitcom or a drama. Did you know that a movie film is really just a string of single pictures flashed one after another so quickly that we think we're actually seeing moving people? Your eyes correctly receive the information about each separate picture, but your brain isn't quick enough to process each picture, so it blurs the information together. You get a moving picture!

You can test this principle by making a *thaumatrope*. A thaumatrope has two pictures, one on each side. When you twirl the thaumatrope, your eyes receive information from each picture faster than your brain can interpret the images. It blurs the two pictures together to form a new and different image. For instance, if there is a bird in one picture and an empty cage in the other, when you twirl the thaumatrope your brain will "see" a bird in a cage! It's not magic—it's seeing! Try it for yourself.

MATERIALS

• pencil
• index card
• markers
• tape

INSTRUCTIONS

Cut an index card in half. On the unruled side, draw two related pictures—one on each piece of the index card. The pictures must make sense together. In other words, draw something like a fish on one side and the ocean on the other side. There is no limit to what you can draw: a bird and a tree branch; a ship and the sea; a table and a bowl of fruit; or even a cat and a ball of string. Tape a pencil to the middle of the back (ruled side) of one of the pieces of index card. Now, tape the other piece of index card to the first so that the pencil is sandwiched between the cards as shown. Now, put the pencil between the palms of your hands and quickly twirl it. You will see the two separate images you drew merge together into one image.

ACROSS

4 This three-pound organ is your electrical control center.

8 This type of light can be dangerous to your skin.

9 You can use this instrument to listen to your heart and lungs.

12 These insects taste with their feet.

14 Your air tube is called a _____.

15 Your _____ transmit electrical impulses that tell your body what to do.

17 This is the strongest muscle in your body.

18 The "rubber floor" of the thoracic cavity.

DOWN

1 This part of your blood carries nutrients to your body cells.

2 This is your body's frame.

3 Your blood carries _____ away from your body cells.

5 You have powerful ones in your legs and arms, but your heart is the strongest _____.

6 Your heart is about the size of your _____.

7 This substance circulates throughout your body.

9 You couldn't taste very well without your sense of _____.

10 An optical toy.

11 These nerves, found high in your nose, are directly linked to your brain.

12 The human body uses food as _____.

13 The cavity in your chest that contains your lungs.

16 Your red blood cells carry this gas to all of your body cells.

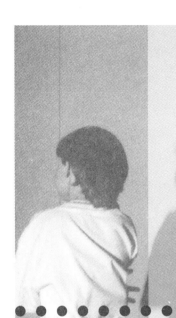

CHAPTER 5

EXPLORING LIGHT

BENDING LIGHT

COLOR

REFLECTION

PINHOLE CAMERA

INTRODUCTION

Light is a very important part of life on Earth. In fact, it very literally allows you to see. It is common sense that you cannot see in the dark. But did you know that when you look at something like a plate of French fries what you are really seeing is light? Packets of energy—called *photons*—stream through the air and bounce off the French fries and the plate and the tablecloth and the bottle of ketchup. . . . What you actually see is this reflected light, which your brain interprets as french fries and a plate and a tablecloth and a bottle of ketchup. . . .

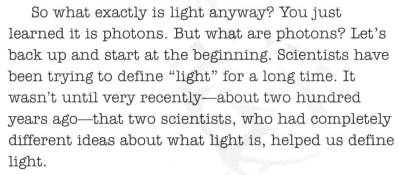

So what exactly is light anyway? You just learned it is photons. But what are photons? Let's back up and start at the beginning. Scientists have been trying to define "light" for a long time. It wasn't until very recently—about two hundred years ago—that two scientists, who had completely different ideas about what light is, helped us define light.

One of these scientists was Sir Isaac Newton. He thought light was made of particles—a stream of invisible particles that resemble tiny balls. The other scientist was Christian Huygens, and he disagreed with Newton. He thought of light as a stream of waves, rather like the ripples that spread out on the surface of a pond when you throw a pebble into the water. And thus began a great scientific debate: Is light a stream of particles or is it a wave? Some scientists agreed with Newton, because his explanation of light could explain how mirrors work. If light were a stream of particles, scientists could understand how these particles bounce off your body, onto a mirror, and back into your eyes so that you could see your reflection in the mirror. But other scientists liked the wave idea because it explained why light comes in different colors. The great debate continued for years as some scientists found reasons why light should be a particle and others found equally valid reasons why it should be a wave.

Then along came a third scientist—Albert Einstein. He discovered the strange nature of light. It is not a particle or a wave, but has properties of both! That might be hard to believe, but it's true. The answer to the particle or wave question is influenced by which question you ask. If you want to describe light bouncing off a mirror, you speak of light as a particle. But if you want to talk about why the sky is blue, you speak of light as a wave. The idea that light has a dual nature is really very simple, but it took a lot of scientists a long time and many experiments to accept it as a fact.

You are probably most familiar with thinking about light as a wave. We speak of many types of light waves—microwaves, X-rays, even radio and television waves. Just like waves in the ocean, these other waves can be fast or slow, wide or narrow. Radio and television waves are big, slow waves. X-rays are short, fast waves. Regular light waves are somewhere in the middle. The range of different kinds of waves is called the *electro-magnetic spectrum.*

Regardless of whether you think of light as a wave or a particle, one thing is sure—light moves fast. Actually, light travels faster than anything else in the universe: 186,000 miles every second! At that speed, the light from the Sun, which is 93 million miles away from the Earth, reaches us in only eight minutes!

BENDING LIGHT

Imagine you are riding your bike on a boat dock that leads down to the ocean. You are picking up speed as you pedal down the ramp, when suddenly you lose your brakes! You zoom right into the water! Besides getting really wet, something very obvious happens to you as you ride into the water—your speed almost immediately slows down. Why? Because water is so much more dense than air; it is simply harder to move through water than it is through air. The same thing happens to light as it hits water. Light zooms through the air at 186,000 miles per second (a lot faster than you can pedal your bike), but if it hits water, it is slowed down. Another thing happens to light as it enters the water— it bends! Actually, light is *refracted*. Light normally obeys Newton's laws and travels in a straight line, but as it moves from one medium to another, it gets displaced—or bent. You can easily see refraction by placing a tilted pencil in a glass half filled with water. Where the light enters the water, the pencil looks broken. Refraction is responsible for this optical illusion.

Light is also refracted when it enters other substances,

such as glass or plastic. If the glass or plastic is curved—as it might be in an eyeglass or telescope lens—the light rays bend even more. If you have ever put on someone else's eyeglasses, you know that the world looks distorted. That is because the lenses are curved in just the right way to bend light specifically for the person who is wearing the eyeglasses. You are also probably familiar with a magnifying glass, which is a *convex* lens, that is, its surface curves outward. When light bends through a *convex* lens, objects seen through that lens appear to be larger than they really are. Just the opposite illusion is created when light is bent by a *concave* lens—a lens with an inward curving surface. In this case, objects appear smaller than they really are.

If you have ever used a magnifying glass, you probably discovered as you looked through it that the world sometimes looks blurry. That happens when the light rays coming through the lens bend and separate, becoming slightly chaotic and disorganized. You may have had to move the magnifying glass forward or backward in order to focus it so that objects became clearly visible. What you were doing—without even knowing it—was adjusting the lens until all the light rays, which were bent as they traveled through the lens, came together in an orderly way. Then the world became crystal clear once again.

MATERIALS

- magnifying glass
- photographic slide
- incandescent desk lamp or flashlight

I-N-S-T-R-U-C-T-I-O-N-S

Dim all the lights in a room that has a white wall. Position and turn on a desk lamp or flashlight about four feet away from the white wall. Hold the photographic slide an inch or two in front of the light source, then place the magnifying glass between the slide and the wall. Move the magnifying glass forward or backward until you find the position where it brings the slide's image on the wall into focus. Now, make some popcorn and invite your family and friends for a grand slide show!

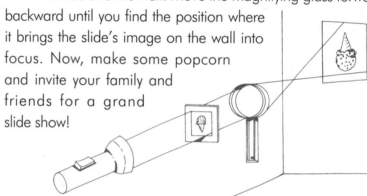

OTHER THINGS TO TRY

Use the procedure above to focus the photographic slide image onto the ceiling.

Look at a friend through a glass of water. He or she will look pretty funny, because the refracted light distorts his or her face. Can you draw a picture of your "refracted friend?"

MUSEUM OF Science.

COLOR

Roses are red, violets are blue. . . . And some slushy drinks are bright green, and a sunset can be purple, and a clown's hair can be orange! But what is a color, and where does it come from? Color is part of plain, old colorless light, like sunlight. Sunlight is technically called *white light,* and it contains all the colors of the rainbow. When light that is all the colors of the rainbow is mixed together, guess what? It makes white light! That's right—white light is actually a mixture of red, orange, yellow, green, blue, indigo, and violet light. Sound crazy? It's not. Just think of a rainbow. A rainbow appears when clouds spread the seemingly invisible and colorless white light of sunlight into its different colors.

Hold on to your hat, because color gets even weirder. Color is a wave! Remember in the chapter introduction we said that light could be thought of as either a stream of particles or a wave? Well, when you are talking about color, it is easier to think of light as a wave. Not like the waves of the ocean, but as invisible waves that travel through the air. All light waves travel at the same speed, but they have other properties that differ—such as *wavelength* and *frequency.*

Color is related to a light ray's *wavelength*. Red, for example, has the longest wavelength; violet has the shortest.

So when you look at a bright red bicycle, you see it as red because, as the light rays strike your bike, all of its color waves except for the red wave are absorbed into the bike. The red wave bounces off the bike and strikes your eye, telling your brain, "This is a red bike." In the same way, your shirt is blue only because it is absorbing all the other color waves and reflecting only the blue wavelength to your eyes. And those black hightops you are wearing—they absorb every color wave so that no color waves bounce back to your eyes. Black is the absorption of all color; white is the reflection of all color.

So your red bike is red because it reflects red light. Does that mean it isn't red in the dark? In a way, that is true. We do not see color well in the dark because there is an absence of light. Another interesting aspect of color is that your bike is red only in white light, like sunlight. If we were to shine a spotlight that had a blue filter on it at your red bike, the two color wavelengths would mix and your bike would look almost black! So it all depends on the light.

Keep your imagination cap on and get ready to experiment with color yourself: Mix different colors together to make white!

MATERIALS

- cardboard
- 3-inch- or 4-inch-diameter round bottle or jar
- scissors
- crayons or markers
- about 2 feet of string

Place the bottle or jar on the cardboard and trace around the bottom to make a disk. Cut out the circle. Now choose crayons for all seven colors of the rainbow: red, orange, yellow, green, blue, indigo (dark purplish blue), and violet. Divide the cardboard into seven equal sections, like the pieces of a pie. Color each section with one of the colors from the rainbow. Poke two small holes in the center of the circle about half an inch apart. Thread the string through the holes and tie the ends together.

Loop the ends of the string over your fingers. Adjust the disk so it is in the middle of the length of string, then twist the disk around until the string has a few good twists in it. Gently pull the loops to get the spinner going. Alternate between gently relaxing the tension on the string and pulling it taut; this will keep the spinner moving. As it twirls, you will see the colors of the rainbow blend together to make the color white!

Try other combinations of colors as well, and try to predict what color you will see! Sometimes cut-glass crystal or diamond rings can break white light into the colors of the rainbow. They act as a *prism*. Ask an adult to help you locate an object that can serve as a prism. Then hold it by a window or other light source and turn it until you see a rainbow!

A *mnemonic* (sounds like ni-mon-ic) is a memory technique that helps you remember information that is long or difficult. For instance, to remember the seven colors of the rainbow, you can create a memorable sentence in which the first letter of every word is the first letter of a color—so that "Rinse Out Your Galoshes Before Inviting Visitors" reminds you of red, orange, yellow, green, blue, indigo, violet. Can you think up a mnemonic sentence of your own?

REFLECTION

Do you remember in the story "Sleeping Beauty" when the evil witch looked into the mirror and asked, "Mirror, Mirror on the wall, who is the fairest of them all?" The mirror told the truth—the fairest person was not the evil witch! In our world, mirrors cannot talk; however, they do "tell" the truth. If you look in the mirror and your hair is sticking up and you have a milk mustache and there are cookie crumbs on your shirt, the mirror will show you just as you are—cookie crumbs and all.

Mirrors simply reflect the light that falls upon their surface. Light usually travels unimpeded through the air, and when it comes into contact with an object, it bounces off that object. When you are standing in front of the mirror, light is bouncing off your body and onto the mirror, then bouncing back off the mirror and into your eyes, where your brain recognizes the person in the mirror as you! The quality of the image reflected in the mirror depends in part on the angle of all the reflected light. If the mirror surface is smooth and flat, then all the light rays that are reflected back from the mirror move together in the same direction, usually straight back at the object. Therefore, the image reflected in the mirror will be clear and undistorted. But if the surface of the mirror is warped, then the light rays don't bounce straight back. Some light rays bounce off at different angles, creating distortions in the image. You can see how angles of reflection distort images by viewing yourself in the crazy mirrors at a circus fun-house. When you stand before one of these mirrors, your reflected image is crazily distorted: you can look as short as an elf or as tall as a giant. Or you can look as skinny as a lollipop stick or as wide as a cow. Some crazy mirrors even make your body look all wavy and rippling. In these cases, the mirror is not really telling the truth. The light creating the image is simply obeying the laws of reflection.

The scientific explanation of the way light angles off a surface—whether it's the mirror in your bathroom or a fun-house mirror—is called the *Law of Reflection*. Basically, it says that the angle at which the light rays hit the surface of an object will determine the angle at which they are reflected back from the object. The study of light and its behavior—including how it angles off objects—is called *optics*. You can thank scientists' knowledge of optics for such inventions as the telescope and the periscope, which both work because of the way light reflects off surfaces.

MATERIALS

- 1-quart paper milk carton, empty and clean
- two 3-inch x 4-inch plastic mirrors (available at local glass shops)
- scissors
- tape

Make sure the milk carton is very clean, because there is nothing worse than a periscope that smells like sour milk! Set the carton upright and ask an adult to help you cut two three-inch slots into one side of the carton—one at the top and one at the bottom—at 45-degree angles to the top and bottom of the carton (see illustration). The slots should be long enough for the mirrors to pass through. Cut identical three-inch slots on the opposite side of the carton. Push one mirror through the bottom slots so that it is inside the carton with its reflective surface facing up. Push the other mirror though the top slots so that it is inside the carton with its reflective surface facing down. Don't push the mirrors all the way into the carton. Keep a small edge on the outside of each side to keep the mirrors from slipping inside the carton. You can secure the mirrors in place by taping these outside edges to the exterior of the carton.

45^0 slot

45^0 slot

Make the spying holes by cutting directly opposite the reflective surface of the mirror at the bottom of the carton. Cut a larger hole (about the same size as the mirror) directly opposite the reflective surface of the mirror at the top of the carton.

Your periscope works because the light coming into its top bounces off the two mirrors and then into your eyes. Now go find something cool to spy on!

Can you design a periscope with a taller scope? How about one that can see around two corners?

• The word "periscope" comes from the Greek: "peri" means "around" and "scope" means "to look around," so periscope means to look around.

• Early mirrors were made by polishing pieces of brass, silver, or gold.

PINHOLE CAMERA

People used to entertain themselves with a gizmo called a *camera obscura* (which means "darkened room"). The camera obscura was a kind of big-screen TV—sometimes it was a roomsize TV! People would pile into a darkened room that had a large screen on a wall, or sometimes in the center of the floor, with the camera obscura located opposite the screen. The camera obscura would project moving images from the scenery outside the room onto the screen. However, there was one big problem with this kind of projector—the images projected were upside down! Not that this detail bothered anybody, because the camera obscura was high technology at that time and people thought it was really cool that they could see any images at all! These days, camera obscuras are hard to find, but you can make your very own miniature version called a "pinhole camera."

To understand how a pinhole camera works, you have to know that light travels in a straight line until it hits an object. Once it hits the object, such as a house, the light reflects off the object's surface. Depending on what the house is made of, the light may bounce off its surface in a fairly focused way, with the light rays all going in roughly the same direction, or in a fairly chaotic way, with light rays scattering in every direction. All that is important for understanding how a pinhole camera works, however, is to recognize that some of that reflected light travels through a tiny hole in the camera and projects an image of the object the light bounced off—a house in this case—onto the back of the camera. Now comes the really interesting part: The image projected onto the back of the camera is upside down! Why? Because light travels in a straight line until it hits an object. Now don't be fooled into thinking a "straight line" means perfectly parallel to the ground or the floor. A straight line can be at an angle from the horizontal, too. So the light reflects off the top of the house; travels in a straight line, but at an angle, through the pinhole in our camera; and hits the bottom of the camera; the light reflected from the bottom of the house travels in a straight line, but at an angle, through the pinhole to the top of the camera. The result is an upside-down house! Build your own pinhole camera and see for yourself!

MATERIALS

- rectangular cardboard box large enough to fit loosely over your head
- tape
- black paint
- white paper
- scissors
- knitting needle
- scarf (optional)

The cardboard box should be roomy enough to fit easily over your head and still have a little extra space. Make sure the box is clean, has all four flaps, and does not have any holes in it. Paint the inside of the box black. Use the white paper as a screen, cutting it to fit one side of the box. Tape it to the inside of the box. Using the knitting needle—with an adult's assistance—make a very small hole, no larger than half an inch in diameter, in the upper corner of the side of the box opposite the screen. It does not matter which upper corner you punch the hole in; you just want to be sure your head will not block the light source. Now cut a hole in the bottom of the box just big enough to put your head through. The camera will work better if no extra light gets into it through this head hole. If necessary, wear a scarf around your neck to block any extra light from entering the pinhole camera.

Now comes the fun part! Go outside on a sunny day and place the pinhole camera on your head. Aim the pinhole at an interesting object and view that object—upside down—inside the box.

ACROSS

1 When you look through this type of camera, you see an upside-down image.
3 Sometimes scientists think about light as a stream of particles, sometimes as a
   _____ .
6 The entire range of different kinds of light waves is called the _____ spectrum.
8 The scientist who thought of light as waves.
9 A lens that makes objects appear larger.
11 The type of lens that makes images appear smaller.
12 The scientist who though of light as a stream of particles.
14 It has seven colors: red, orange, yellow, green, blue, indigo, and violet.
16 "Rinse Out Your Golashes Before Inviting Visitors" is an example of this kind of memory technique.

DOWN

1 Light particles are called _____ .
2 A Greek word meaning "around."
3 When all of the colors of the spectrum are reflected, light is this color.
4 When light hits a mirror, it is
   _____ .
5 When light travels through water, it is bent, or
   _____ .
7 The study of light and its behavior.
10 The scientist who discovered that light is neither a particle nor a wave, but has properties of both.
13 A piece of glass or plastic that can break light into the colors of the rainbow.
15 The absorption of all light makes
   _____ .

OTHER THINGS TO TRY

Cut another hole in the bottom of the pinhole camera that you can fit your hand through, so that your hand can reach the screen. Tape a smaller piece of white paper on the screen and project an image onto this paper. Now, trace the image. This method of reproducing an image may not be quite the same as capturing an image on film, but you can still draw an amazingly realistic image.

F·U·N F·A·C·T·S

Your eyes work just like a pinhole camera—images that are reflected onto the back of your eye are upside down! The reason you don't see the world upside down is that your brain "interprets" the light image and turns it right-side up!

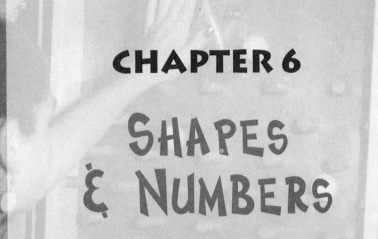

CHAPTER 6

SHAPES & NUMBERS

MOBIUS STRIP
TANGRAMS
PROBABILITY
TESSELLATIONS

INTRODUCTION

What do telling time, counting out the correct change in order to buy a pack of baseball cards, and rearranging your bedroom furniture all have in common? Give up? They all require that you know mathematics! Telling time and counting change require that you know such math basics as adding, subtracting, multiplying, and dividing. Rearranging furniture means you must have a sense of distances and widths and how shapes fit together. All of these skills are math skills. And you thought math was only about memorizing times tables and suffering through quizzes at school!

The truth is, math is not just a bunch of formulas or tables to be memorized; it is the study of patterns and is a way of thinking about a problem in order to arrive at a solution that can be verified and repeated, which is what science is all about. Suppose your dad has baked a huge sheet cake for your birthday and your 30 hungry friends each want a piece of that cake. What's the best way to cut the cake so everyone gets an equal piece? Arriving at a solution to that problem involves thinking about how shapes can be manipulated and divided. Is it better to cut squares, rectangles, or perhaps even thin strips? This problem involves *geometry*—the mathematical study of shapes, lines, solids, and other forms.

Now suppose you have to play detective because the last Popsicle is missing from the freezer. You go through the list of possible suspects: Your four-year-old brother loves Popsicles, but he's too small to reach the freezer compartment. Your mom is away on a business trip, so she couldn't have eaten it. Unless there is a Popsicle burglar loose in the neighborhood, there is only one likely suspect left—your dad. You have arrived at a possible solution through a process called *logic,* which is an important principle in math and in detective work!

Knowing about math can be useful in many ways. Suppose your mother insists that you use a new toothpaste—let's call it Brand X—but you want to stick with your bubble gum-flavored fluoride toothpaste. Your mother says, "But it says on the label that Brand X is recommended by nine out of ten dentists. It's good for you!" Your mother, believe it or not, is using a branch of mathematics called *statistics* to try to convince you to switch toothpastes! You can tell her to beware, however, because statistics can be very deceptive. For instance, if the toothpaste package does not tell you how many dentists were surveyed, the "nine out of ten" statistic will not have very much meaning. After all, what if the toothpaste manufacturer asked only ten dentists? That means that thousands of other dentists could have hated the toothpaste! "Nine out of ten" is not very meaningful unless you know how many dentists were surveyed, which is called the *sample size.*

Here's one more example of how important it is to understand math. Tell your parents you have a great deal for them: You will do the dinner dishes each night if they will pay you a reasonable "salary." Your compensation will be one penny the first night, two pennies the second, four the third night, eight the fourth night, and so on, doubling the amount of pennies each night. Sounds like a deal, doesn't it? But figure out what your parents will be paying you after one month. They may change their minds when they see the answer!

We use math every day of our lives in hundreds of obvious and subtle ways. It should be no surprise, then, that all of science is based on mathematics. Math is called the "language" of science. With mathematics, scientists from different countries, who may speak different languages or study different scientific disciplines—such as physics, engineering, or biology—can all speak to each other. Furthermore, scientists must know math in order to carry out experiments. If they did not know math, how would they be able to perform such basic functions as measuring chemicals, reading the temperature on a thermometer, or creating a graph to show their results?

But math is more than just a language used as a tool for other activities. Math is an activity in itself. Just as writers use the language of words to create works of literature, mathematicians use the language of numbers and formulas to make their own creations. As you will see in this chapter, math is not only important in our lives, it is also fun. Go ahead and try the activities on the following pages. We promise there will be no quizzes!

MOBIUS STRIP

To a topologist, there is no difference between a chocolate-frosted doughnut and the cup of milk you dunk it in. Well, there's a big difference in the way they taste, but there is no difference in how their surfaces can be studied. You see, a *topologist* is a mathematician who studies the properties of shapes, the way one shape can be twisted, stretched, bent, and squashed to form another shape. Topologists are interested to know how one shape is the same as or different from another shape. It is the shape of your doughnut and the shape of the mug holding the milk that fascinates topologists. These shapes are the "same" in that they both have an "inside" and an "outside." In this way they are no different from, say, the letters "O" and "D," which also have an inside and an outside. Not all shapes do. For instance, a shape called a *torus*, which is like a doughnut with a twist in it, has no inside or outside.

To a topologist, you are shaped more like a torus than a doughnut, because you have no inside or outside—well, at least your stomach cannot be said to be "inside" your body. Sounds crazy, doesn't it? But if you think of your body as a shape, then it makes sense that the placement of your digestive system is a matter for debate. Your stomach, after all, is connected to the outside of your body through tubes—on one end by your throat, which contains a tube called an esophagus that leads from your mouth to your stomach, and on the other end by your anus. A topologist examining you as a shape could take a marker pen and draw one continuous line from your mouth, down your esophagus to your stomach, through your intestines, out your anus, and continue all the way up your body to your mouth again. In other words, topologically speaking, your digestive system is not really inside your body at all!

Your bones are another story. They are "inside" your body, according to topologists. Remember that chocolate-frosted doughnut? Topologists say it has two distinct surfaces—we'll call them an inside and an outside—because the circle you could trace around the outside surface of the doughnut cannot be connected to the circle you could draw around the perimeter of the inside hole. To connect them, you have to cross one of the lines or tunnel through the doughnut. The two lines of the doughnut cannot be connected in one continuous line, as your digestive system could be. So, there are two separate surfaces on a doughnut. Your bones are like the doughnut, topologically. They are truly inside your body, because you have to tunnel into your body to connect the surface of your body with your bones.

So what is topology useful for? Well, it helps scientists understand the shapes and behavior of DNA molecules, which are the building blocks of life. And this helps biologists understand the patterns of protein in organisms and develop methods for mapping genes. Astronomers and physicists who study topology may be better able to explain the way certain structures in the universe might have formed.

Don't worry if you have a little trouble understanding these concepts or visualizing these shapes. Topologists do, too! That is why, whenever they can, they make three-dimensional or computer models of the shapes they are exploring. Follow the instructions below to make models of some interesting shapes that will help you understand the principles of topology.

MATERIALS

- several strips of paper 2 inches wide and 1 ½ feet to 2 feet long (adding-machine paper works well)
- tape or glue stick
- scissors
- pencil
- notebook

Glue the ends of one strip of paper together to make a huge loop. With a pencil, draw a line in the middle of the strip all the way around the inside surface of the paper. Draw a second line around the outside surface of the same strip. You can easily see two separate lines—this shape has a definite inside and outside. Topologically, this loop is a shape with two surfaces.

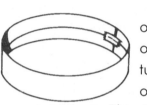

Now take another strip of paper, and lay it out flat on the floor or on a table. Pick up one end of the strip and twist it once by turning it over. Glue this end to the other end of the strip. You should have a loop with one twist in it. This is called a *Mobius strip*. Now draw a line down the middle of the strip on its inside. You'll soon find that you have traced one long line, which connects back to the place you started drawing the line from! The Mobius strip has no clear inside or outside. According to topology, it is a shape with one surface.

Now, experiment further with this strange shape. Cut the Mobius strip along the line you drew down the middle of the strip. What do you think you will end up with? Two separate loops? Think again! Make another Mobius strip, and instead of cutting it down the middle, try cutting it closer to the edge, about one-third of the way in. Do you think you will end up with the same shape you did when you cut the Mobius strip down the middle? If not, why not?

Try making several different loops with different numbers of twists in them. What happens when you draw a line down the middle of a strip with two twists? When you cut that strip along the line? Follow the same steps for a loop with three twists, with four twists. Keep a written record of the number of twists and what happens when you cut along the line. Can you find a pattern? From that pattern, can you guess what will happen if you draw a center line and cut a loop from a strip with 125 twists?

TANGRAMS

Long ago in China, there lived a man named Tan. He had many beautiful possessions, but the one he most treasured was a square clay tile decorated with a scene from nature. Tan admired the beauty of this tile so much that he often carried the tile with him. One day, when Tan had grown old and his eyesight was failing, he was carrying this tile on his daily walk when he tripped over the root of a tree. His precious tile flew out of his hands and smashed on a nearby rock. Saddened, Tan ran over to where the tile lay in seven pieces. He looked at the pieces, and although they were lovely shapes (five pieces were shaped like triangles, one was a square, and the other was a parallelogram—a slanted rectangle), he shed tears to see his beautiful tile broken. He sat with the fragments for a long time, until gradually his tears stopped, for he began to see the beauty of the individual pieces. He realized he was actually a very fortunate man: Now he had seven beautiful pieces of tile instead of just one! Tan spent the remainder of his days admiring and studying these lovely shapes, fitting them into different pleasing arrangements. To this day, the unique patterns created by these seven tile shapes are called *Tangrams*, named after the old man who so loved them.

The details of this story may have been exaggerated over the centuries, but scientists do believe that the Tangram shapes, or something very similar to them, were created in China about four thousand years ago. Ever since, all kinds of people have had fun creating the multitude of patterns that can be formed from these shapes.

Tangrams fall under the branch of mathematics called *geometry*, which is the study of shapes and the patterns of shapes. Specifically, Tangrams explore *spatial relationships.* That sounds like an intimidating term, but it is nothing more than seeing how shapes fit into different spaces. For example, you have to judge spatial relationships to see whether that new computer desk will fit into your bedroom. If you've ever tried to cram a juicy pear into an already packed lunch box, you have played with spatial relationships!

Tangrams are a fun way to study spatial relationships because the seven tiles can fit together—following certain ground rules, of course!—in many creative, and challenging, ways.

MATERIALS

- cardboard
- photocopy of Tangram template
- 8 ½-inch x 11-inch piece of paper
- Tangram template from page 288

I·N·S·T·R·U·C·T·I·O·N·S

Photocopy or trace the shapes in this booklet onto a piece of cardboard and cut them out.

Select one of the large triangles to use as a base shape. Now select other shapes that will exactly fit over and cover up this triangle without any of the edges hanging over or any piece overlapping another piece. Hint: There are two solutions to this puzzle. Try the same experiment using the medium-size triangle as the base shape.

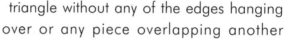

How many different squares can you make out of the seven pieces? Now try to make the shapes illustrated. Then, create your own shapes. Why not trace the outline of an original shape you have created on a piece of paper? Give the tracing to a friend and challenge him or her to fill it in with the Tangram pieces!

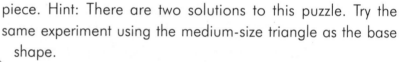

PROBABILITY

There is only one chance in a million that you will be struck by lightning. There is about one chance in ten million that you will be hit on the head by a piece of an airplane that's fallen off during flight. The chance is only one in three hundred million that you will ever be bitten by a shark! How, you ask, do we know any of these things? Because of a branch of mathematics called *probability*. Probability is the likelihood that a particular event will occur.

Experts can figure the probability for just about any event—from the chance that you will win the lottery to the likelihood that you will hit a hole in one next time you golf! They can also determine the probability of more serious events, such as the probability that your house will burn down or that an earthquake will occur in a particular region. Insurance companies rely on these probabilities to determine how much they will charge customers to insure their homes. When accountants help people establish a retirement plan, they use probabilities to determine the likelihood of a person living to a certain age. Some people bet on horse races according to the "odds" that a particular horse will win the race. "Odds" is just another way of saying "probability."

If they wanted to, your teachers could use probability mathematics to figure out what grade you are likely to score on your next test! How? Well, let's figure that out. Suppose you have received a B on every test so far this year. The *likelihood* is that you will continue this trend, scoring a B on the next test as well. Does this mean you definitely will get a B? No way! Remember, a probability is just the *likelihood* of a certain event occurring. If you study hard, you might earn an A. Or if you do not study at all, you might score an F!

You have probably figured out probabilities yourself. If you have ever flipped a coin and called "heads" or "tails," you have entered the land of probability! What do you think the probability is that a coin will show heads rather than tails on any particular flip? Will heads show most of the time? None of the time? Half of the time? Experiment to see for yourself. Toss a coin twenty times and keep a tally of how many times "heads" shows and how many times "tails" shows. Is there any pattern? How about if you continue keeping a record for one hundred flips of the coin?

You can also explore probability very easily using dice. In fact, do that now by having a dice-powered "turtle" race!

MATERIALS

- pair of dice
- piece of lined paper
- pencil

Pretend you are at the edge of a pond at a turtle race track. Twelve turtles are about to race. You want to purchase the turtle who races the fastest so you can take it home and run it in a local race. Because you are a savvy shopper, you do not want to buy any old turtle. You want a top—notch racer. But how will you know which turtle to buy? One way is to watch the race and determine which turtle has the best probability of winning on your home track.

KEEPING SCORE

Set up a score sheet by numbering the lines down the paper 1 through 12. Line 1 is for turtle 1, line 2 for turtle 2, and so on.

THE RACE

Begin the race by rolling the dice and determining their sum (add together the numbers that show on each die face). A turtle "moves forward" each time its number comes up. For instance, if the die faces are showing a 1 and a 4 (adding up to 5), mark an X on line 5. This means turtle 5 has moved forward. If the die faces each come up with a 6 (adding up to 12), mark an X on line 12. Roll the dice 50 times, recording how each turtle moved. Do you have any guesses as to which turtle will move the fastest? Hint: The answer has to do with figuring the probability of which number comes up most often when two die are rolled.) Once you have recorded 50 rolls, add up the number of Xs on each line of the score sheet to see which turtle moved the fastest (has the most number of Xs).

SOME THINGS TO THINK ABOUT
Why do you suppose one number came up significantly more times than the other numbers? Why did one number never come up at all? Think about the way each number combination from 1 to 12 can come up on two die faces.

TESSELLATIONS

You are not going to believe this is a math activity. You will probably think we are trying to sneak an art activity into this book, and maybe we are! Math and art have a lot in common. Throughout the centuries, mathematicians and scientists of all kinds have appreciated how beautiful the mathematical expressions of nature are, just as artists have appreciated the beauty of color and shape. Actually, you already know that math has a lot to do with spaces and shapes. Math is not so different from art after all! Just as art uses patterns, perhaps of stripes (lines) and dots (circles), math uses patterns. For instance, take at look at this pattern: $5 \times 1 = 5$; $5 \times 2 = 10$; $5 \times 3 = 15$; $5 \times 4 = 20$. You may have noticed that the last digit in each answer is either a 5 or a 0, and that the appearance of these digits alternates: 5, 0, 5, 0. That pattern is not an accident; it helps mathematicians discover "laws" about math, meaning that they can find out why math works so well.

Not only numbers follow patterns; shapes do as well. As we have already pointed out, the study of shapes is called *geometry*. Mathematicians have found many patterns in different kinds of shapes. For example, a square has four sides and four corners, but a triangle has three sides and three corners. Looks like a pattern is emerging! If a shape called a hexagon has six sides, can you guess how many corners it has? How about an octagon—if it has eight sides, how many corners does it have? If you said 6 and 8, respectively, you correctly detected the geometric pattern in these shapes. Good for you!

Shapes themselves can be fit together to form patterns. These repeating patterns are called *tessellations*. There are only a few rules about how a tessellation can be formed. Two of the most important are that the shapes that make up the pattern must fit together with no gaps, and they cannot overlap. Chances are you can see examples of tessellations right now. Just look around. If there is a tile floor, a brick wall, a shingled roof, or a chain link fence within view, then you are seeing tessellations. Can you see anything else around you that has a tessellation pattern?

Let's combine art and math right now to make our own creative tessellations.

MATERIALS

- 2-inch-square piece of cardboard
- scissors
- tape
- pencil
- markers or crayons
- drawing paper

Draw a line—squiggly, angular, or curving—that starts at one corner of the cardboard, extends out toward the middle, and ends at the other corner of the same side. Cut along this line and then move the cutout pattern so that its left edge is flush with the right edge of the square. Tape the pattern in place (see illustration). Then trace the shape you have created onto a large piece of paper and make it tessellate by tracing the pattern over and over with no gaps and no overlaps.

Now it gets really fun! Using the same piece of cardboard, draw another line across the bottom, reaching from corner to corner. Remember, this line can be squiggly or straight. Cut along this line, move the cutout so that its bottom edge is seated against the top edge of the original shape, and tape the cutout in place. Now make it tessellate: Trace the new shape onto a piece of paper as you did for the first shape, and see how the

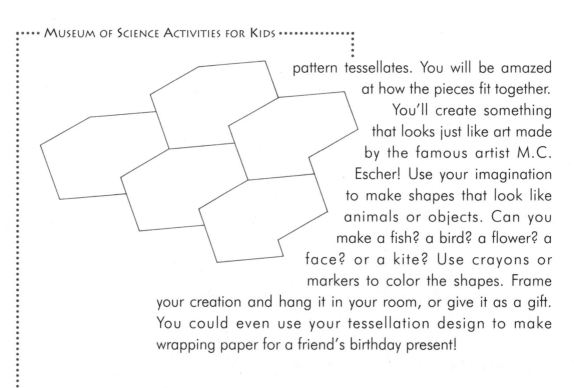

pattern tessellates. You will be amazed at how the pieces fit together. You'll create something that looks just like art made by the famous artist M.C. Escher! Use your imagination to make shapes that look like animals or objects. Can you make a fish? a bird? a flower? a face? or a kite? Use crayons or markers to color the shapes. Frame your creation and hang it in your room, or give it as a gift. You could even use your tessellation design to make wrapping paper for a friend's birthday present!

Go to your local library and see whether they have any books by the artist M. C. Escher. He was a master at creating tessellations. Try your hand at creating some tessellation artwork like Escher's.

The word "tessellation" comes from the Latin word *tessella*, which means "small, square stone." Throughout the centuries, artists have pieced together colorful square tiles to create beautiful tessellations called *mosaics*.

ACROSS

2 The likelihood that a particular event will occur.

4 When the same shape can be fitted together to form repeating patterns, these patterns are called _____.

8 The mathematical study of shapes, lines, solids, and other forms.

11 How many sides does a Mobius strip have?

12 "_____" is just another way of saying "probability."

13 The branch of mathematics that we use to interpret surveys and their results.

14 The name of the set of seven shapes created in China about four thousand years ago.

DOWN

1 One important principle in both mathematics and detective work is _____.

3 A topologist would say you are shaped like one of these.

5 The artist who was a master at creating tessellations.

6 Tangrams explore _____ relationships because they let you see how different shapes fit together in space.

7 Math is called the "_____" of science.

9 A one-sided surface that you can make by twisting a strip of paper and taping the ends together.

10 A mathematician who studies the properties of shapes.

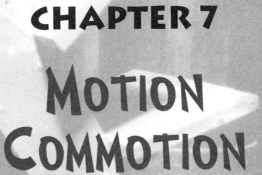

CHAPTER 7

MOTION COMMOTION

• •

BALLOON ROCKET
MAGIC FRICTION BALL
FINGER POOL

INTRODUCTION

Suppose that on the way to your room to change into your rollerblades, you kick your sneakers off in the middle of the living room floor, and then take off, leaving them there. Not that you would do such a thing—but just suppose you did. They sit there all day and all night. The next day your dad reminds you that those sneakers are not going to get up and walk into your closet by themselves, so would you please put them away. But somehow you get sidetracked and forget. And you forget the next day, and the next. . . . Imagine you forgot for an entire month! Your sneakers would just sit there, in the middle of the living room, undisturbed. What do you suppose would happen if they sat there untouched for a year? For fifty years? Would your sneakers still be sitting in the middle of the living room floor if no one touched them for one hundred years? You bet they would! Your sneakers are an example of an important principle of motion called the *first law of motion* or the *law of inertia*. This law states that objects do not change their movement unless a force is applied to them. Or as scientists phrase the first law of motion: Objects at rest will remain at rest and objects in motion will remain in motion, unless they are acted upon by an outside force. In other words, your sneakers are not going anywhere unless you—or someone or something else—move them.

It is common sense that objects at rest (like your sneakers) stay at rest. However, the opposite is also true: Objects that are moving tend to stay in motion, unless affected by an outside force. This principle may seem a little harder to believe. If you think of Earth orbiting the Sun, it may seem reasonable that Earth will keep moving forever, but what about objects on Earth? After all, when you run, you are able to stop yourself. And everyone knows a ball won't keep rolling forever. The reason objects on Earth eventually stop moving, however, is that there are forces that slow them down. For instance, a bowling ball stops rolling because the bowling alley it is rolling down affects it through a force called *friction*. Friction is the force that causes resistance between moving objects that are in contact with one another. The microscopic bumps on the wood surface of the bowling alley and on the surface of the bowling ball rub against each other, causing the bowling ball to slow down and eventually stop. The more bumps, the more resistance, and thus the more friction. If the bowling alley were made of gravel, friction would be increased and the bowling ball would stop rolling much more quickly. In contrast, if the bowling alley were covered with snail slime, friction would be reduced and the ball would continue rolling for a much longer time. However, you probably would not want to use it again! There are other forces at work slowing the bowling ball down—such as air resistance and gravity—but you get the general idea about how force affects an object's motion.

There are two more laws of motion. Don't worry—if you break them you won't be arrested! Actually, it is impossible to break the laws of nature. The second law of motion says two things: The more force exerted on an object, the more the object accelerates; and the heavier an object is, the more force it will take to accelerate it. Remember those sneakers sitting in the middle of the living room floor? Nudge one with your toe, and it will probably move a couple of inches. But kick it— apply more force—and it will fly across the room! To experiment with the second part of the second law, kick a heavier, more massive object. Imagine kicking the refrigerator. It doesn't take much imagination to see that it takes a lot more muscle power to move a refrigerator than it does to move a sneaker.

The *third law of motion* says that nothing can change its own movement without acting on, or being acted on, by something else. What does this mean? Well, think about walking. When you walk across the floor, you're pushing against the floor, which moves a little but not much—because it's attached to the rest of the building. Since the floor doesn't move much, it pushes back on you and you take a step forward. Here's where friction comes into play again. You need the friction between your feet and the floor to walk. If this sounds strange, try walking on ice. There's very little friction. You can't push very hard on the ice, because if you do, your feet go one way and the rest of your body goes the other. You're not walking; you're falling down!

The laws of motion are not man-made laws. It's not like at school—if you break rules, you might end up in detention hall! The laws of motion are physical laws, and they define the way everything in the universe works.

BALLOON ROCKET

People get pushed around every day—not by bullies, but by everything around them in the world! If you are sitting in a chair, the chair is pushing against you and you are pushing against the chair. The reason you stay seated is that the forces are equal, so they cancel each other out. If the forces were unequal, movement would result. For instance, if the chair exerted a far greater force on your body than your body exerted on the chair, you would go sailing into orbit! All this pushing is explained by the third law of motion.

Try it yourself! It's easy to see the third law in action, especially if you have a friend about the same size as you are and you're both wearing roller skates! Put on your skates and position yourselves close together and opposite each other. Then, at the same time, reach out and gently push each other apart. If you each use the same amount of force in the push, then you will both roll apart an equal distance.

Now, if your father showed up—on roller skates—and took your friend's place in this experiment, you can be sure that even if you each pushed with the same force, your dad probably wouldn't travel the same distance you did. Why?

Because he's bigger and heavier than you are. *Gravity,* which is a force of attraction, pulls more on him than it does on you. The more force gravity exerts on his body, the greater the friction between his skates and the floor surface, and the more easily his motion is slowed down. Because he's so big and weighs so much, you have to exert more force—more muscle power—to move him than he has to exert to move you.

Now, what if some of your friends came over to assist you with these motion experiments? They can be a big help, because if you push your dad as a group, you will discover that he is a lot easier to move than when you're pushing him by yourself. That's because the group exerts more force than you can all by yourself. If this scenario really took place, then your dad would help demonstrate the second law of motion: The greater the force applied to an object, the greater the object's acceleration, and the greater the mass of the object, the greater the force needed to accelerate the object. Acceleration isn't just speeding up—it is a change in speed or in direction over time.

These laws of motion apply to everything that moves—from your dad to a NASA rocket. However, it is easier to see these laws in action in a rocket. So instead of analyzing your dad's motion, let's build a rocket to examine the laws of motion. A rocket sits unmoving on the launching pad until its engines fire. This demonstrates the first law of motion, which says that an object at rest tends to stay at rest until some force is applied to it. The engines blast hot gases with great force against the ground, propelling the rocket upward (the third law). The rocket weighs tons, so the engines must be very powerful and produce a lot of force to get the rocket moving (the second law). It's unlikely NASA will invite you to use their facilities to build a rocket for your experiment, so why not build a balloon rocket in your own room?

MATERIALS

- balloon
- 10 feet of string
- drinking straw
- tape

SET-UP

Tie a string to the back of a chair or to some other stationary object. Thread a drinking straw onto the string. Tape the still-deflated balloon to the side of the straw so that the mouth of the balloon is facing away from the chair. Then pull the straw down to the free end of the string.

BLAST OFF

Blow up the balloon and pinch off the mouth so no air will escape. Don't tie the balloon's mouth closed! Now pull the string so it is taut, and let the balloon go. It will shoot along the string. Blast off!

• Set up two strings parallel to each other and have balloon rocket races with a friend. Experiment to see what will make your balloon travel faster. Try more air in the balloon, smoother string, differently shaped ballons, etc.

• With a brightly colored felt-tip marker, mark the string about halfway to the chair. Can you control the motion of your balloon rocket so that it stops at the mark?

• Rockets were used seven hundred years ago in China. They were powered by gunpowder and used in military defense.

• Some race cars are propelled by rockets.

MAGIC FRICTION BALL

Roll a marble across your living room floor: Where will it stop? Does it keep rolling at the same rate of speed across the room until it hits a wall? Or does it roll partway across the room, slowing down and eventually stopping, usually well short of the other side of the room? Your experience is probably that the marble stops well short of the opposite wall. But have you ever stopped to think why? If there were no walls to stop it, why wouldn't a marble just keep rolling forever? The reason is a force called *friction*. Or at least, friction is most of the reason. There are a lot of forces affecting the marble and how it rolls, but friction is one of the most important. Friction is a force that causes resistance between objects that come into contact, especially moving objects. Friction slows objects down. Try rubbing your hands together vigorously. They heat up and feel kind of sticky, don't they? That's friction! Friction causes the same kind of "stickiness" between the surfaces of objects. For instance, a wooden floor may appear smooth, but in reality it is covered with microscopic grooves and bumps. So is the surface of a marble, and just about any other surface you can think of.

147

Even glass and ice do, although they have far smaller bumps than a wooden floor. The bumps on the floor and the bumps of the marble rub against each other, and because of the friction this rubbing creates, the rolling marble slows down and eventually stops. Because ice has smaller bumps, it has less friction. A grassy field, on the other hand, has a lot of big bumps and so would create a lot of friction. So a marble would roll farther across ice than it would roll across grass.

Friction has its advantages and its disadvantages. One advantage is that it helps us, and most other animals, to walk! It is friction that keeps us from slipping. If we were to walk across a newly polished floor, we'd be much less likely to slip if we were wearing sneakers than if we were wearing loafers. Why? Because the grooved rubber soles of sneakers are designed with good gripping power to take advantage of friction. The slick leather surfaces of loafers are not. There are animals who have feet as well designed for not slipping as sneakers! One example is a lizard called a gecko. Its feet are covered with tiny ridges that fit into the microscopic bumps and crevices of the surfaces of rocks, trees, and even glass! As a result, a gecko can walk securely just about anywhere—even upside down on a ceiling! A polar bear's feet use friction to their advantage as well. Friction is created when the rough skin and tough hair on their feet move against the smooth ice. Scientists have even studied polar bear feet in an effort to make better boots, so we humans can walk on ice without slipping!

But friction also has its disadvantages. For instance, friction can cause objects to wear away. Remember the heat generated as the friction built up while you were rubbing your hands together? That heat and the wear-and-tear on the surface of moving objects as they experience friction can cause mechanical parts to wear out or break. Think of a car's engine. It would work a lot better if less friction were generated by all its moving parts. Oil, grease, and other lubricants are added or applied to car parts in order to keep their surfaces from rubbing so hard and creating so much friction. Even air can cause friction! Airplanes experience friction as they fly through the air. The airplane bumps into the air molecules—nitrogen, oxygen, and carbon dioxide molecules—which slow the plane down slightly. Airplanes shapes are no accident. They are designed to be as streamlined as possible so that they will move through the air making as little contact with air molecules as possible. Dolphins have the same "aerodynamic" shape so that they can swim through water molecules with as little friction as possible.

Despite our serious discussion, we shouldn't forget that friction can also be fun! Magicians use friction to create many of their illusions. Whether it's using friction to "stick" a

MATERIALS

- 3 feet of aluminum foil
- pen or other small tubular object
- 3 feet of string
- tape
- tweezers (optional)

coin in a place that makes it look like it has disappeared, or to manipulate how the cards in a deck are shuffled, magicians are masters of using friction to their advantage and the audience's disadvantage. You can experiment with friction by doing magic, too. Follow the directions below to "magically" control a ball of aluminum foil with nothing but your voice!

I-N-S-T-R-U-C-T-I-O-N-S

Scrunch the aluminum foil into a loose ball. Use the pen to poke a tunnel at an angle halfway through the ball of aluminum foil. Create another angled tunnel on the other side of the aluminum ball, directly opposite the first tunnel. The two tunnels should meet at the center of the aluminum ball, creating one V-shaped tunnel.

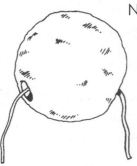

Next, push the string through the V-shaped tunnel. If you need to stiffen the string to thread it through the ball, try tightly wrapping some tape around the end (rather like a shoelace tip) that is fed through the tunnel. If necessary, use tweezers to draw the string through the tunnel.

Once the string is through the ball, position the ball close to one end of the string. Now, hold the string vertically and pull it taut; that is, stretch the string tight, with one hand holding the end nearest the ball up and the other hand pulling the bottom end of the string down. Now, let the string relax a little. As you give the string some slack, the foil ball will slide down the string. When you pull the string taut, the ball will stop sliding downward. Why? Because of friction! As the string becomes taut, friction increases as the surface of the string and the surface of the foil ball that is touching the string rub together. Because of the increased friction, the ball won't slide over the string as easily and will come to a stop.

Where's the magic? Try this. (Like all good magicians, you will want to practice this trick before actually trying it out on your family and friends.) As you loosen the string, yell "Go!" The ball will begin to slip down the string, looking like it is following your command. Then yell "Stop!" as you pull the string taut. The ball will magically stop on your command!

FINGER POOL

What do a teacup sitting on a cupboard shelf and a baseball being hit out of the ballpark have in common? *Momentum!* What is that, you are probably asking. It is a force that acts upon all objects. Technically, it is the measure of an object's motion that is equal to its *mass* (how much material the object has—similar to its weight) multiplied by its *velocity* (the object's speed in a particular direction). But let's not get too technical! You can get a good idea of what momentum is by comparing how hard it is to stop the motion of two different objects. Suppose a Ping-Pong ball were sliding down a slide. It would not take much energy for you to reach out and stop the ball's slide. Now, suppose a bowling ball were sliding down the slide. It would take a lot more effort on your part to stop the bowling ball. You could say the Ping-Pong ball—which has relatively little mass and velocity—also has little momentum. The bowling ball—which has greater mass and velocity—has more momentum.

And what about the teacup and the baseball? Because neither the teacup nor the cupboard shelf is in motion, their momentum equals zero. Actually, there are forces at work on both the teacup and the shelf. The teacup is pushing down on the shelf, and the shelf is pushing up on the teacup, but because the forces are equal, they cancel each other out and the teacup stays put. The baseball and bat push on each other, too, but with unequal force. In the split-second collision when the bat hits the ball, some of the bat's momentum—which is generally greater than the baseball's—is transferred to the ball. The ball's momentum increases so that, the batter hopes, it will sail right out of the ballpark!

Momentum always travels from one moving object to another when the objects make contact. Momentum is never "lost." For instance, as the baseball sails through the air, it begins to slow down; it loses some of its momentum. One of the reasons this happens is that the ball is colliding with air molecules. It transfers some of its momentum to the air molecules, which gain momentum themselves and bump into each other. Eventually, because of gravity, the ball is drawn toward the ground, where it transfers the rest of its momentum to the ground as it lands. Since Earth is so huge, it absorbs the momentum without a noticeable shake. If the baseball landed on a car windshield, however, how do you think the windshield glass would be affected by the ball's momentum? You guessed it—it would probably crack or shatter. But the glass's momentum would still be transferred to the objects it was in contact with, like the air and the car itself. So, you see, momentum is never lost; it just keeps going forever! Scientists call this fact the "conservation of momentum."

MATERIALS

- rectangular cardboard box
- sharp knife (Adult Assistance Required)
- 5 quarters
- marker pen
- ruler

(Adult Assistance Needed)

The surface of a pool table, whether it is felt-covered slate or cardboard, should be as flat and smooth as possible. Therefore, choose the smoothest, flattest side of the cardboard box to be your pool table surface. Ask an adult to cut four small, square holes—each large enough for a quarter to fit through—at each corner of the "pool table." Now you are ready to play finger pool!

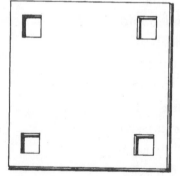

FINGER POOL OBJECTIVE

Finger pool is best played with two players. The object of the game is to be the first person to sink two quarters in the corner pockets.

FINGER POOL PRACTICE EXERCISES

All good pool players spend a lot of time practicing. To get a feel for how the quarters move on a cardboard pool table, try the following exercises. Before you get started, designate one of the quarters as the cue ball by coloring it with a marker pen.

Exercise 1

Flick the cue ball with your fingers so that it slides across the pool table and hits a quarter. When you flick the cue ball, you are adding momentum to it, so that it goes from a resting position to a moving one. When the cue ball hits a quarter, it transfers some of its momentum to that quarter, which then moves across the table. Experiment with flicking the cue ball with different amounts of force.

Exercise 2

Line three quarters in a row. Flick the cue ball head on into the first quarter in the row and watch what happens to the last quarter in the row. It moves! The momentum of the cue ball is transferred to the first quarter, which transfers some of its momentum to the second quarter, which in turn transfers some of its momentum to the third quarter. The momentum moves through the row so that hitting only the first quarter in the row makes the last quarter move!

Finger Pool Playing Rules

Set four quarters and a "cue ball" on the pool table. Group the four quarters in a square in the middle of the table, two of them with tails up and two with heads up.

ACROSS

2 In the finger pool activity, you transfer
_____ from one quarter to another.

4 The force that causes resistance between moving
objects that are in contact with each other.

6 More than just speeding up, this is a change in
speed or direction over time.

9 The "_____ of momentum"
describes the fact that momentum is never lost,
it just passes from one object to another.

10 The _____ law of motion says that
objects at rest will remain at rest and objects in
motion will remain in motion unless acted on by
an outside force.

11 When you walk, you push against the floor,
which pushes back against your feet, enabling
you to move. This demonstrates which law of
motion?

12 The laws of motion are not man-made laws.
They are _____ laws and define the
way everything in the universe works.

DOWN

1 Oil, grease, and other _____ are
applied to car parts in order to keep their
surfaces from creating friction.

3 The scientist who discovered the laws of motion.

5 The _____ law of motion says that
the more force exerted on an object, the more
the object accelerates; and the heavier an object
is, the more force it will take to accelerate it.

7 Rockets powered by gunpowder were used seven
hundred years ago in this country.

8 The tendency of an object to remain at rest or in
motion (also the first law of motion).

Players take turns flicking the cue ball at the quarters in an effort to sink them into the corner pockets. One player must sink only the "heads" quarters and the other player must sink the "tails" quarters.

SOMETHING TO THINK ABOUT

Because momentum is neither created nor destroyed, where does it come from, and where does it go when all the quarters stop moving?

CHAPTER 8

SIMPLE MACHINES

INTRODUCTION

What do you think of when you hear the word "machine?" Noisy equipment at a factory? How about farm tractors, jack hammers, cranes? Perhaps you think of household machines, such as washing machines, toasters, or vacuum cleaners. We can guess that you did not think of scissors, tweezers, nutcrackers, or screws. But these are machines! You see, a machine is any device that helps you accomplish a task—to do work—with less effort. It does not take a lot of imagination to see that washing clothes by hand takes more effort than tossing them in a machine that washes them for you. It is easier to crack a nut with a nutcracker than with your bare hand. And a crane can certainly lift heavy objects that even twenty people working together couldn't lift. Now, you may still be wondering about the screw. How can a screw be a machine? The threads of a screw enable it to do work while the screw turns. For instance, as you turn a screw, it is able to "pull itself" into a piece of wood, metal, or other material.

There are two ways to think about, or categorize, machines: as simple and as complex. Now *simple* does not mean "easy to understand,"

and *complex* does not mean "hard to understand." These terms simply mean that complex machines are made up of a lot of simple machines. Think of "simple" as "basic"—things like gears, screws, pulleys, and levers that perform basic actions like turning, pushing, pulling, and lifting. A screwdriver is a simple machine because it is little more than a wheel and axle used to increase the force of your hand. In contrast, a complex machine is one that has different basic machines combined within it. A vacuum cleaner, for example, is a complex machine because it contains rollers, a belt, wheels, and a motor that contains gears.

All machines reduce the amount of work we have to do or the effort we have to exert to do that work. When we speak of work, we are referring to the physical effort it takes to accomplish a task, such as lifting a motorbike onto the back of a truck so you can transport it to a safe track to ride it. You would probably use a ramp to help you get the motorbike onto the truck bed, right? Well, the ramp is a machine because it reduces the effort you need to exert to get the bike onto the truck!

Work can be broken down into two parts. One part is the amount of *force* (what we referred to as "effort" earlier) needed to accomplish the task. The other part is *distance,* or how far the force will be spread. Distance is important to understanding how much force is needed to accomplish a job, because the smaller the distance, the greater the force needed to get the job done. Conversely, the longer the distance, the less force is needed to accomplish the job. It takes less force to overcome inertia over a long distance than it does over a short distance. *Inertia* is a law of physics that tells us that objects at rest tend to stay at rest until affected by a force, or objects in motion tend to stay in motion until affected by a force. Think back to that motorbike that needs to be loaded onto the truck. There is very little distance between the bike on the ground and the truck bed. So a lot of force will be needed to lift the bike off the ground—to overcome its resting inertia—and onto the truck. You would probably need help to accomplish this task. But add a ramp and you add distance between the bike and the truck bed. It takes less force to get the bike moving, rolling up the ramp, and then onto the truck than it does to lift it straight up. You could probably roll the bike up the ramp and onto the truck bed all by yourself. As the slope of the ramp changes, the effort needed to move the bike also changes. A steep ramp requires more effort than one with a gentle slope. Ramps helped the ancient Egyptians and Meso-Americans use other simple machines to haul huge stones into place to build pyramids.

You use simple and complex machines all day without thinking about them—or, perhaps, appreciating them-but the minute you no longer have them—yikes! From doorknobs to drills, from staplers to steering wheels, from hockey sticks to helicopters, machines make our lives easier and more enjoyable.

PINWHEEL AND AXLE

A pinwheel is not only a lot of fun, it is a simple machine. That's right! A pinwheel is what is called a *wheel-and-axle machine*. Just as its name says, a wheel-and-axle machine is made up of a wheel and of a shaft called an *axle*. The two parts are combined in such a way that when one part turns, the other does as well. The most common example of a wheel and axle is the steering wheel of a car. The steering wheel is attached to an axle that moves a series of parts connected to the wheels. When you turn the steering wheel to the left, the wheels turn to the left, and the car turns left.

Another example of a wheel and axle is a *windlass*. Windlasses used to be common in the days when water had to be drawn from wells. You can still see these simple machines in action on some farms or in a Western movie. It's the windlass that you crank to lower and raise the water bucket. The windlass has a crank that is attached to an axle. The bucket and rope are then attached to the axle. So when you turn the crank, the axle rotates and unwinds the rope, lowering the bucket into the well. Reverse the direction of the crank, and the axle reverses direction, winding the rope and lifting the water bucket. Ahhh, a cool drink!

If you have ever tried hauling water this way, you may have noticed that the larger the crank, the easier it is to raise the water bucket. The smaller the crank, the more work it takes to raise the bucket. Why? Because a small crank turns in a small circle and turns the axle only a little bit at a time. Therefore, it takes more force to crank the bucket up. A big crank, however, turns in a big circle, covering more distance. The axle turns a lot more quickly, and so it is easier to haul the bucket of water up.

Can you think of a way not to have to turn that crank at all? Perhaps the movement of a river could push the crank in circles (as it does in a water wheel). Or maybe even the force of the wind could do the work for you. Let's check that principle out by making a pinwheel, which, as we said before, is nothing more than a wheel-and-axle machine!

MATERIALS

- paper
- tape
- scissors
- plastic drinking straw
- thread
- paper clip
- a quarter
- ballpoint pen or a single-hole punch

Cut an eight-by-eight-inch square from a piece of paper. Fold the lower-left corner up to the upper-right corner. The paper will be shaped in a triangle. Make a strong crease and then unfold the paper. This time fold the lower-right corner to the upper-left corner. Crease the paper and then unfold it again. You now have a square with creases in it forming an X. Next, place a dime in the center of the X and trace around it to mark a circle. Remove the dime, and punch a hole with a ballpoint pen or a hole puncher in the center of the circle. Then punch a hole at each corner of the square as shown. (The holes should be located to the left of the crease line in the top two corners and to the right of the crease line in the bottom two corners.) Next, cut along the creases that form the X, but stop just before you reach the circle made by the dime. Then, fold each of the four corners containing the holes into the center, aligning all the holes. Insert the straw through the holes in the pinwheel and push the

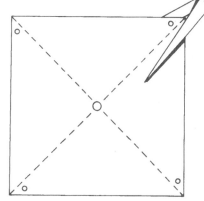

pinwheel halfway down the straw. Wrap some tape on the straw on either side of the pinwheel to keep the pinwheel from sliding on the straw. Now, cut a length of thread about twelve inches long. Tape one end to the straw toward one end and attach the paper clip to the other end of the thrend.

You are now ready to put your wheel-and-axle machine to work hauling the paper clip up and down. Hold the straw in both hands as follows: Hold your hands with your palms facing each other or facing you—which ever is more comfortable. Lay the straw loosely between each thumb and index finger, as shown in the illustration. Make sure the string with the paper clip is hanging freely. Then, blow on the pinwheel. As the pinwheel turns, it turns the straw axle. The axle motion causes the thread to wind up, hauling up the paper clip.

Make a larger pinwheel. Is it easier to wind up the paper clip?

BLOCK AND TACKLE

Superman could leap tall buildings and lift heavy objects without breaking a sweat, but do you know you can too? Well, you might not be able to leap buildings, but you could probably lift a heavy object, such as a refrigerator, if you needed to. How? With a simple machine called a *block and tackle*. A block and tackle is a collection of two or more pulleys. A *pulley* is a kind of wheel-and-axle machine that is used to lift objects or move them. Mechanics use a block and tackle to lift a car engine out of a car; on certain kinds of boats, sailors use it to lift sails to the top of the mast; cable cars in San Francisco are pulled along by an underground pulley system. A pulley has a wheel that turns on an axle within a housing. The pulley wheel has a groove along its outer edge, in which fits a rope or cable. You attach whatever object you want to lift to one end of the rope, and then you pull on the other end of the rope.

However, one pulley alone doesn't reduce your effort, it just changes the direction of effort. Using a pulley may only *seem* to make the job easier, because it is usually easier to pull an object than it is to lift it straight up. But you are still exerting almost as much force—in fact even a little more force, because of friction. The real energy savings comes when you combine more than one pulley into a block-and-tackle machine. Then it really does take less effort to raise an object. See, to lift heavy objects you don't have to be superhuman after all!

MATERIALS

- 2 empty thread spools
- 7 feet of string
- 1 empty half-gallon plastic milk or detergent jugs with handles and caps
- 2 wire coat hangers
- wire cutters
- pliers
- a strong horizontal bar (shower curtain rod or a broom balanced between two chairs)

(Adult Assistance Needed)
MAKING A PULLEY

Ask an adult for assistance cutting a twelve-inch section of a wire coat hanger. Once the piece is cut, slip a spool onto the wire so that about four inches of wire stick out of the spool on one end and about seven inches stick out from the other. Use the pliers to bend the coat hanger up about a quarter of an inch from where the wire comes out of the spool. The wire is your axle. The spool is the pulley wheel.

Next, use the pliers to loop the four-inch end of the wire around the seven-inch end and secure it there. Then, form the free end of the wire into a hook. You have now made a pulley! However, as we said, combining pulleys into a block and tackle adds lifting power, so go ahead and make another pulley right now.

Using the Pulley

Hook one pulley over the bar, the shower curtain rod, or broom handle. We'll call this pulley the fixed pulley, because it will stay in place. (The other pulley will be able to move. But it comes later!) Loop a piece of string over the spool of the fixed pulley. Now, fill the plastic jug with water and cap it. Tie one end of the string of the fixed pulley to the handle of the plastic jug, and take the other end of the string in your hand. Pull down. As you do, the jug should begin to rise.

Now, add a second pulley to make a block-and-tackle machine, and compare how much effort it takes to raise the water-filled jug. First, you need to untie the string from the jug and unthread it from the fixed-pulley wheel. Once that is done, hook the second pulley to the handle of the jug; this is the moveable pulley. Now, tie one end of the string directly to the bar (shower curtain rod or broom handle). Then, loop the other end of the string under the spool of the moveable pulley (which will hang upside down) and up over the wheel of the fixed pulley. Your block and tackle is ready to go to work! Pull down on the string to raise the jug. Is it easier to raise the jug with a block and tackle than it was with a single pulley?

In order to lift extremely heavy items, such as a refrigerator, you would have to run the rope through the block-and-tackle system several times. This is because increasing the distance will allow you to do more work without applying more force. You can try this on a smaller scale by running the string through the block and tackle as described above, but looping it one more time through the moveable and fixed pulleys. Notice how much easier it is to lift the water-filled jug. You'll also notice that you have to pull a heck of a lot more string through the block and tackle to raise the jug just a little bit off the ground! Remember what we said in the chapter introduction about the greater the distance, the less force it takes to move an object? Well, that is the principle at work here. Less force means spreading the effort out over a greater distance, which means more string!

PANTOGRAPH

Have you ever doodled on a piece of paper and really liked the picture you drew? Perhaps you decided to redraw it larger so you could pin it on your bedroom wall. But you could never quite get the larger picture to look like the smaller one. Or perhaps you saw a photograph you liked and tried to draw it poster size. Again, the larger version just did not look as good as the original. Well, don't despair— simple machines to the rescue! By making a *pantograph*, you can copy smaller drawings or photographs in order to make larger versions that look like the original. And this is all made possible because of a simple machine called a *lever*.

A lever is a very simple machine that is used to multiply force. It is usually a long arm-like object used in conjunction with a *fulcrum*, a fixed-pivot point on which the lever sits. Levers are in use all around you: A seesaw in a playground and a hockey stick are nothing more than big levers. The lever arm of a seesaw is the plank on whose ends the two people sit. The fulcrum is the middle base of the teeter-totter on which the lever arm rests. How about a hockey stick— where is its arm and fulcrum? Give up? The lever arm is the

entire hockey stick and the fulcrum is your hand!

Levers are usually used to lift heavy objects. But don't take our word for it; try this experiment to test out the lever's excellent ability to lift objects. Position a chair so its back is to the edge of a table. Remove everything breakable from the tabletop. Now, place a broom handle over the back of the chair and under the edge of the table. Gently push down on the broom where it sticks out toward you over the chair back. What happens? The table is very easily raised up! The effort, or force, you applied to the broom handle lever moved the load, which is the table. The chair acted as the fulcrum. Repeat this experiment after moving the chair out from the table a little farther. Is it easier or harder to raise the table now? How does the position of the fulcrum change the amount of effort that is needed to move the load?

Seesaws and the chair-and-table lever you created above are called *first-class levers*, because the fulcrum is located between the effort and the load. There is another kind of lever called a *second-class lever*; its fulcrum is positioned at one end of the lever arm, and the effort is at the other end. Think of a wheelbarrow. The wheel at the front end of the wheelbarrow is the fulcrum; the effort is at the end with the handles, where you push the wheelbarrow. The load is in the middle. It takes more effort to lift a load using a second-class lever than using a first-class lever. And a third-class lever requires even more effort. In fact a third-class lever is used not to make a load easier to lift, but to make that object move farther. That is because in a third-class lever, the fulcrum is at one end, the load is at the other, but the effort is in the middle! Ice tongs are an example of a third-class lever. The fulcrum is at one end, where the two lever arms come together. The load—the ice cube—is at the other end. The effort—where your hands squeeze the lever arms—comes in the middle. Ice is not heavy, so the lever does not help you lift it; but ice is slippery and too cold to hold, so the lever allows you to move it easily. Talk about a simple machine!

So what has all this got to do with drawing pictures? Well, you can make lots of really neat toys that use levers. A pantograph is one of them. Using its lever arm, you can make perfect enlargements of a smaller drawing or photograph. Go ahead and make one, and see if you aren't soon creating masterpieces!

MATERIALS

- rigid corrugated cardboard, cut to the following lengths:
  two 2" x 9" strips
  two 2" x 5" strips
  six 2" x 2" squares
- a large 2' x 2' piece of sturdy cardboard
- 2 pencils
- scissors
- a nail or other sharp object to punch holes
- drawing paper
- tracing paper

I-N-S-T-R-U-C-T-I-O-N-S

(Adult Assistance Needed)

Ask an adult to help you punch holes in the strips at the points indicated by the black dots in the illustration. Arrange the two 2-by-9-inch strips and the two 2-by-5-inch strips as shown. Fasten the strips together with paper fasteners so that you make a small square in the corner. Now take the six 2-inch squares and ask an adult to help you punch holes in their centers. Divide the squares into two stacks of three pieces each. Glue each stack together by spreading a layer of craft glue on both sides of the piece that will be in the middle and then sandwiching it between the two outer pieces. These stacks will be pencil braces. When the glue has dried, glue one stack to the enlarging end and one stack to the reducing end. You have made a pantograph, which is a type of lever used for enlarging and reducing.

Before you can begin enlarging or reducing, ask an adult to help you punch a hole in the cardboard square at the lower left corner and secure the pantograph to this square of cardboard at hole #3 using a paper fastener. This will be your drawing board. When you attach the pantograph, position it so the fastener heads are in contact with the paper. This will make the pantograph slide smoothly on the paper. Slip a sheet of drawing paper between the pantograph and the cardboard sheet. Then, insert a pencil in the pencil brace located at hole #1. It should be in an upright position, with its point touching the paper. Finally, hold a pencil in hole #2 and draw a shape. The pantograph lever moves with your hand, so the other pencil will draw the exact same shape as the one you are holding, only it will be larger! If you switch positions and trace with the pencil in hole #1 and the tracer in hole #2, the drawing will be smaller.

Now, remove the paper and put in a sheet of tracing paper. Insert a photo or a picture of something you want to draw underneath the tracing paper. Then trace the picture or photo with the pencil in hole #2, but don't really let this pencil point touch the paper. Only pretend to trace it. You'll see why in a moment. As you move the pantograph arm over the contours of the picture you are "tracing," the pencil in hole #1, which is touching the paper, will be drawing an enlarged version of the picture!

F-U-N F-A-C-T-s

Archimedes was a Greek mathematician and inventor who lived more than two thousand years ago. He said that if he had a long enough lever arm and a strong enough fulcrum, he could lift the world? Do you think he could have?

OTHER THINGS TO TRY

Examine the pantograph and identify the lever arms, the effort, the fulcrum, and the load. Can you tell what type of levers these are?

MECHANICAL SAILING SHIP

What goes round and round and up and down? The answer could be a rollercoaster, a merry-go-round, or a crankshaft. A crankshaft is a simple machine that can be combined with a wheel-and-axle machine to make a lot of very fun toys. It may not sound as much fun as a merry-go-round, but it is one of the reasons a merry-go-round works! You see, a *crankshaft* is a rod or shaft to which other rods are attached. When the main rod moves, the other rods move and also move anything that is attached to them. Think of those horses on the merry-go-round. They move up and down on crankshafts. Or the needle on a sewing machine. It is raised and lowered on a tiny crankshaft. Now, when a crankshaft is combined with a wheel-and-axle machine—which we hope you will remember is a wheel that turns a shaft as it moves—something strange happens: circular motion, or *rotational motion,* is changed into linear motion, also called *reciprocating motion,* which is movement either back and forth or up and down. But this transformation is not magic. It is *mechanics.*

Let's make a mechanical toy that demonstrates the interesting things that can result from combining a crankshaft with a wheel-and-axle machine. Let's make a sailing ship that can ride the roughest of seas!

MATERIALS

- coat hanger
- wire cutters
- empty cereal box
- plastic drinking straw
- 2 pencils
- 2 paper fasteners
- paper
- cardboard or card stock (about 2 square inches)
- tape
- scissors
- ruler

(Adult Assistance Needed)

Get an adult's assistance to open and straighten a wire hanger and to cut a sixteen-inch section from it. If the wire cutters do not cut through the wire, try bending the wire back and forth at the point where you want to cut it. Sometimes the bending motion itself will cause the wire to snap. Then, use the wire cutters to bend the wire into the shape illustrated.

To prepare the cereal box, tape the top flaps closed. Then, turn the box upside down. Because the bottom of a cereal box is sturdier than the top, we will use it as our top. It will be referred to as "the top" from here on! The bent wire is your crankshaft. Lay the crankshaft wire across the top of the box to check that the ends protrude over the ends of the box. If they don't, you have cut your wire too short! If your wire is the correct length, hold it in place on the box top; use a pencil to make two spots on the box top exactly in the middle of each of the two bends, called

CRANK HANDLE

CRANK ARM
←1 INCH→

←3 INCHES→   x   ←2 INCHES→   x   ←2 INCHES→

←1 INCH→

CRANK ARM

←1 INCH→
←2 INCHES→
←1 INCH→

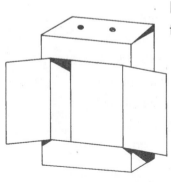

the crank arms, in the wire (see illustration). Now, remove the wire and, with an adult's assistance, punch two holes at these spots. Make sure the holes are big enough for a pencil to fit through.

Next, get an adult's assistance to punch a hole on each side of the box about six inches down from the top. Also get assistance to cut two doors into the front of the box. Bend the doors open.

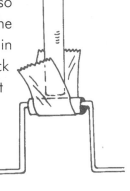

Now, cut two sections from the drinking straw, each about three quarters of an inch long. Slide them over the crankshaft wire until they are seated in each of the bends, or crank arms, of the wire. Make sure the straw sections will spin freely on the crank arms. If not, trim the straw sections.

It's time now to fit the crankshaft into the box. Open the doors on the front of the box and insert the crankshaft so that the ends stick out through the holes in the sides of the box. Bend the wire that protrudes from the left side down in a 90-degree angle so that the crankshaft will not slip back into the box. The right side will serve as the crankshaft handle. Finally, push a pencil through each of the holes in the top of the box. They should touch the crankshaft at the crank arms, where the straw sections are. If they don't, adjust the wire. Cut two long pieces of tape and have someone hold the pencils against the straw

ACROSS

3 A wind-powered, wheel-and-axle machine that you can make with a straw and a piece of folded paper.

7 A screwdriver is an example of a wheel and _____ machine.

8 Another word for "linear motion."

10 The physical effort it takes to accomplish a task.

12 A fixed pivot on which a lever sits.

13 This machine uses levers to help you copy a picture and make it bigger at the same time.

15 How many classes of levers are there?

16 A rod or shaft to which other rods are attached.

17 Another word for "circular motion."

DOWN

1 Simple machines work by increasing the _____ over which your effort or force must be spread.

2 You can find this first-class lever at the playground.

4 Ice tongs are an example of which class of lever?

5 A very simple machine used to multiply force. It is usually a long, arm-like object that is used in conjunction with a fulcrum.

6 A simple machine can decrease this by increasing distance.

9 A block and _____ is a collection of two or more pulleys.

10 A wheel-and-axle machine used to draw water from a well.

11 A type of wheel-and-axle machine that has a wheel with a groove along its outer edge.

14 Any device that helps you to do work with less effort.

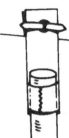

sections while you tape the pencils to the straws. When you are done, make sure the straw sections still move freely on the crank arms. Close the doors on the front of the box.

Tape a thin strip of cardboard about an inch long to the end of each pencil where it protrudes from the box. Now, draw a wild pirate ship or whatever kind of ship you want. Cut out your ship and attach it to the cardboard on the ends of the two pencils with paper fasteners as shown. Also, draw some waves and tape them to the front of the cereal box so it looks like the ship is sailing through the ocean. Now, crank the crankshaft handle and watch your ship toss and turn over the waves! (Hint: If the box tends to topple over from all the motion, put something heavy inside it to weigh it down.)

CHAPTER 9

Noisy Fun

PITCH PIPES
....
STRING TELEPHONE
....
SENSATIONAL SONIC EARS
....
MUSICAL INSTRUMENTS
....

INTRODUCTION

Have you ever heard the sound of silence?

Probably not, because it is almost impossible to experience absolute silence in our everyday lives. However, we hear the sounds of what we think is "silence" all the time. Even when you are the only one home and you are sitting "silently," chances are that there are sounds you can hear: a refrigerator motor, the leaves rustling in the trees in your backyard, your own heart beating. Just about the only natural place you can hear absolute silence is in outer space! Outer space has no atmosphere, so sound cannot travel. It is absolutely silent. You see, sound needs a "medium" to travel through, like water or air. Here on Earth, sound travels very easily, so our world is full of it. You really cannot get away from sound.

Sound is a shaky science. What we mean is that wherever there is sound, there is a whole lot of shaking going on! Shaking as in *vibration,* which is a quivering motion that usually occurs when something moves very quickly up or down, or forward and backward. Stretch a rubber band between your fingers and pluck it. The rubber will vibrate, causing a twanging sound. Place your fingers gently on your throat and hum your favorite song. Can you feel the vibrations of your vocal cords—vibrations that make the sound of your voice? Your voice is actually the sound of air passing over the vocal cords (also called your voice box) in your throat. The air movement makes them vibrate, and the way you control the air movement through your mouth and nose shapes the sound of the vibration into words or humming or a yell or a groan.

Okay, so now you know how sound is created, but you may be asking, How do I hear it? Good question! When objects vibrate, they cause other objects nearby to begin vibrating as well. For example, when you plucked that stretched rubber band, the vibration of the band set the air molecules around it vibrating. Now, air molecules are very tiny, invisible objects, but there are billions of them in every cubic inch of air, so they are all around you. When a vibrating object sets the air molecules around it vibrating, the air molecules in turn set their neighbors vibrating too, which creates a growing wave of vibrations. The wave's motion is kind of like when you are in line at school—perhaps waiting to go out to recess—and the last person in line shoves into the person in front of him or her. Then that person falls into the person next in line, and that person falls into the next person, until the whole line is moving forward. That's a wave of motion! To get technical, when the wave of motion brings objects closer together, scientists call the motion *compression,* and when the motion causes objects to separate, the motion is called *rarefaction.*

Now, all those vibrating air molecules create a wave that finally reaches your ear. If your ears work well, the vibration that was carried through the air collides with your eardrum and you hear sound. We will talk more about how your ears work later, but for now let's just concentrate on sound. You may think it would take a long time for sound to reach your ears. But sound waves are speedsters: They travel through air at more than a thousand feet per second. That means sound travels about one mile in just five seconds!

Sound waves travel even faster through liquids such as water and solids such as stone or steel. They travel faster through these denser mediums because the wave becomes more compressed. In air, the wave is able to spread out; so one molecule takes its time "bumping" its neighbor and moving the wave forward. But in dense objects, like water and rock, the wave is compressed, so the "bumping" happens quickly and the wave moves fast. Sound travels through water at about 5,000 feet per second, and it can travel a very long way before the wave loses its strength and the sound fades away. This is why whales can communicate over such long distances. Sound travels through solid steel at the amazing speed of about 20,000 feet per second!

It may seem that loud sounds travel faster than quieter sounds. For example, your dad's voice yelling "Turn that music down!" seems to travel a lot faster than his voice does when he says "Please take out the garbage." But the truth is, loud and soft sounds travel at the same speed. Loud sounds just have more energy—called *intensity*—than quieter sounds. The loudness of sound is measured in *decibels*, a term that means the measure is in units of ten ("deci"), and honors an important researcher of sound, Alexander Graham *Bell*. Hence, the word "decibel." The decibel range is a series of numbers, and the higher the number assigned to a sound, the louder the sound. Zero decibels equals the softest sound humans can hear. Leaves rustling in a tree measure about 20 decibels, whereas a crowd of people conversing measures about 60 decibels. A live rock concert measures about 100 decibels, and a thunderstorm about 120 decibels.

The rock concert measure may not seem to be much greater than the loudness of the crowd of people talking, but consider this: For every ten decibels you add, the level of loudness goes up ten times. So, for example, the rock concert (100 decibels) is ten times louder than heavy auto traffic at 90 decibels. Listening to any sound that is 100 decibels or louder for any sustained period of time can permanently damage your hearing. That is why people who continually work around loud noises (at airports, construction sites, or rock concerts) must wear ear protection.

PITCH PIPES

Here is a silly joke for you: Why did Tim bring a ladder to music class? Give up? Because his teacher wanted him to sing higher!

What Tim's teacher really wanted was not for Tim to elevate his body, but for him to sing in a higher-pitched voice. Anyone who can sing, no matter how badly, can change the pitch of his or her voice, from low to high. *Pitch* describes how high or how low a sound is. A clown, for instance, may speak in a high, squeaky voice. A tuba makes a low, deep sound. It is very easy to distinguish the pitch of these two sounds. Experiment for a minute with your own voice: How low and how high can you sing?

Scientists have developed a very accurate way of determining the pitch of sounds using a measurement called a *hertz* (named after the man who invented it). A hertz measures how fast or slow a sound wave vibrates. One hertz is equal to one vibration per second. The high, squeaky clown's voice would register a lot of hertz because the higher the sound, the faster the sound wave is vibrating and the higher the hertz measurement. Just the opposite is true for low sounds: They vibrate slowly, so a tuba would have a low hertz measurement. And what about you? The human voice can make sounds that measure anywhere from 85 hertz to 1,100 hertz. That's a wide range of sound! But some animals can top humans—bats, porpoises, and grasshoppers, for example, can make sounds that are more than 100,000 hertz! Because our ear can only hear sound up to 20,000 hertz, that means there are a lot of sounds in the world—including bat, grasshopper, and porpoise calls—that we are deaf to.

Life would be pretty boring if we could hear only one pitch. Think of a robot talking in a *monotone,* which means it talks all in one pitch. Fortunately for us, we can hear a lot of different pitches within our hearing range. That means we can appreciate beautiful sounds such as music, which uses variations in pitch to create complex but pleasing sounds. In fact, music is based on a *scale* of sounds that move from a low pitch to a high pitch in regular increments. The word "scale" comes from the Latin word for "staircase," which gives you some idea of the orderliness of the pitches of a musical scale.

You can experiment with musical sounds and the musical scale by playing a pitch pipe. Pitch pipes are small, round, flute-like objects on which you can blow each individual pitch in the musical scale. Musicians use pitch pipes to tune their instruments or decide the pitch for a song they want to sing.

You can make a pitch pipe using a bottle and water! You just fill a bottle with water to a certain height and blow across the mouth of the bottle to produce a sound. How does it work? Well, there is air inside the bottle, so as you blow across the mouth of the bottle, you set the air inside the bottle vibrating. If there is only a little water in the bottle and, thus, a lot of air, the air molecules vibrate slowly (because they have a lot of room to move around, and don't bump into each other as often). This slow vibration means the sound will be low pitched. If there is a lot of water and less air, however, the sound will be high pitched. So as you vary the amount of water, the pitch of the sound changes. Follow the instructions below to make and experiment with a water-bottle pitch pipe.

MATERIALS

- 5 small empty soda bottles, preferably glass, all the same size
- pencil
- paper
- tape
- water

Cut the paper into five labels, one for each bottle. Mark the labels "A," "B," "C," "D," and "E." Tape one of the labels on each bottle. Then place the bottles in a row in order from "A" to "E." Each bottle now represents a note on the musical scale! Next, carefully pour the following amounts of water into each bottle: one inch of water into A, two inches into B, three inches into C, four inches into D, and five inches into E. Then pucker up

and blow! Well, it may take you a little while to perfect your musical ability—or even to produce a sound! Practice by blowing gently across the mouth of a bottle. Here's a good technique to try: Place your bottom lip gently against the lower lip of the soda bottle opening, then blow across the surface of the opening, not down into the bottle. Once you have mastered the blowing technique, try to play the musical scale by blowing on each bottle in order from A to E. You will hear the pitch change from low to high.

Ready to make some music? Try to "play" the following "notes" by blowing on the bottles labeled with the letter that matches the note. A space in the notes below means to leave a slight pause. Can you figure out what song you are playing?

CBAB CCC
BBB CDD
CBAB CCC
CBB CBA

Try making up your own song, and write out the notes so your friends can learn to play it too!

OTHER THINGS TO TRY

Tap the sides of the soda bottles with a wooden spoon. What happens? Does the sound change? Do you know why? (Hint: The water vibrates instead of the air to produce sound.) Now try making other kinds of pitch pipes. Fill bottles with sand, marbles, or paper instead of water. What kind of sounds can you make?

STRING TELEPHONE

Can you imagine a world without telephones? How would you make plans with your friends, alert the fire department in an emergency, or order a pizza? Life without telephones is pretty difficult to imagine, isn't it? So it's surprising to think they are a relatively recent invention: They were invented only slightly more than one hundred years ago. Now, one hundred years may seem like forever to you, but it is really not that long ago. It is very likely that your grandparents remember getting their first telephone. That phone may have been attached to a line that many people shared, so your grandfather may have had to make sure someone else wasn't talking before he attempted to make a call. Or he may not have been able to dial a person direct. Instead, he may have had to ring the operator to route his call through to the right person. It wasn't so easy using a telephone back in the "olden" days!

The telephone was invented in 1876, by Alexander Graham Bell. We have him to thank for starting the telephone revolution, which has resulted in cellular and car telephones, faxes, and computer modems. Those first phones were really not so different from today's phones, because all phones transmit sound vibrations. This is how the process works: You pick up the phone and dial the pizza place. When someone answers, you ask for a large mushroom pizza. Now, the sound waves you emit from your mouth are nothing more than air molecules vibrating. The air vibrates over your vocal cords, which "shape" the sound into the words "one mushroom pizza for delivery, please." The sound waves continue to travel—right into the phone receiver. Inside the receiver is a mechanism that changes air vibrations into an electronic signal, which travels through the telephone wires to the phone assigned to the number you dialed—in this case, the pizza shop. When the signal gets to that phone, a mechanism in the ear piece there translates the electronic impulses back into sound waves. The pizza person hears your voice and takes your order!

It is not necessary to change sound waves into electronic signals for them to travel down a wire. You can attach two strings to a cup and make a crude telephone. String phones work because sound vibrations are pretty good at traveling down string. In fact, they travel through solid objects, such as string, better than they do through air, because the molecules making up the solid object are packed closer together than air molecules are. You can prove this to yourself by tapping your finger on a table and listening for the quality of the sound. Now tap the table again, but press your ear to the tabletop while you tap. Doesn't the tap sound louder? Try placing something that ticks, like a loud clock or an oven timer, on the table. Does it sound louder when you are just listening to it as usual, or when you are listening to it with your ear pressed to the table? Well, it is time to get your ear off the table and onto the telephone receiver! Let's make a few phone calls: string-phone calls!

MATERIALS

- 4 to 6 feet of string
- 2 paper or plastic cups
- 2 paper clips
- scissors
- a partner

INSTRUCTIONS

Carefully poke a small hole in the bottom center of each cup. Thread the string through the holes and tie a paper clip to each end so they won't slip back through the holes. Now have your partner take one cup while you take the other. Stretch the string out between you. First, you talk (into your cup) while your partner listens (by placing the cup over his or her ear). Then, let your partner talk while you listen.

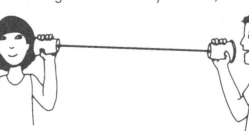

That's all it takes to make a phone: a couple of cups and some string for the sound waves to travel along. This may be one phone your parents don't yell at you to get off!

The quality of the sound of your string phone will vary depending upon the type of cups and string you use to make it. Besides regular paper or plastic drinking cups, try tin cans, yogurt containers, or other kinds of "cups." Instead of household string, try kite string, yarn, jute, or even wire.

Make a four-way phone by linking two string phones together!

If there's no one else around to talk with on your string phone, you can see how sound travels on a string with a wire coat hanger. Tie one end of the string to the coat hanger and wind the other end around your finger. Very gently put that finger into your ear, and let the coat hanger dangle in the air. Now bang the wire hanger against something solid, such as the wall or a chair (be careful you don't damage the wall or chair!), and then let the coat hanger dangle in the air. You will hear an amazing symphony of sound travel from the coat hanger, down the string, and into your ear!

SENSATIONAL SONIC EARS

Elephants have huge ears. Lizards have tiny ears, which are really just tiny holes in the sides of their heads that are covered by scales. Compared with elephant and lizard ears, human ears are . . . what? Medium-size? How about just about the right size! As you know, a human ear is a hole in the side of the head that is surrounded by a fleshy, folded disk of skin and cartilage that is called the *auricle*. Now that we know what human ears look like, it is time to describe how they work!

Remember how earlier in this chapter we described that sound waves are just air molecules vibrating? Well, ears are designed specifically to detect and channel those sound waves. The auricle captures sound waves and funnels them into the hole in our head, which is called the *ear canal*. The sound waves travel down the ear canal until they bump into a tiny structure called the *eardrum*. Now, our eardrum does not look like a drum, but it acts like one! The vibrating air molecules of the sound waves strike the eardrum and start it vibrating, which starts a chain reaction of vibrations. The eardrum vibration sets tiny bones in the ear vibrating, and these set tiny fluid-filled tubes in our ears vibrating, and these tubes start tiny hairs in our ears vibrating. Here things get a little more complicated. When the tiny hairs vibrate, they change the physical vibrations into electrical signals, which they send on to our brain. The brain receives these signals and "interprets" them as a specific sound—a rock song, your mother's voice, a balloon popping! Incredible, huh?

Almost all animals hear because of this same process. So why, you might ask, are the appearances of their ears so different? The reason has a lot to do with their environment. For example, a fox has large moveable auricles because if it wants to eat, it has to be able to funnel the greatest amount of sound possible, even the quietest noises, into its ears. That's because foxes love to eat mice, which are rather quiet creatures. Other animals, such as birds, have no auricles. Birds' ears are just little holes in their head covered with feathers. Flat ears make their bodies more streamlined, which is necessary for flying.

There are times when we all wish we could hear better (like when someone is telling a secret!). You can actually help yourself hear a little better if you cup your hand around your auricle. That way, you capture more sound and funnel it into your ear. You might find yourself doing that naturally when you are at a noisy event and are trying to hear someone talk to you. Make the sensational sonic ears described below, and you may never have a problem hearing again!

Before you begin, think about what shape of outer ear will likely help you hear better. Perhaps you will want to sketch a few designs before you actually make your sonic ears. Consider that the greater the size of the auricle, the more sound is likely to get funneled into your ear canal.

Once you've designed a pair of ears (auricles, really), use the plastic tubs to make them by cutting and taping pieces together. Hold the new ears against your own ears. Now, test your sensational sonic ears. You will need a partner who will drop a penny on a noncarpeted floor. Before you begin, stand across the room from the person with the penny and note your exact position by putting a piece of masking tape on the floor. Close your eyes. Now, have your partner drop the penny. Did you hear it? If so, take five giant steps farther away from your partner. Mark your new position with another piece of masking tape. Now, close your eyes and listen for the penny again. If you heard it, repeat this procedure until you can no longer hear the penny drop. Now, take off your sensational sonic ears and repeat this entire experiment using only your real ears. Were you able to get as far away from your partner as

MATERIALS

- 2 clean plastic containers (margarine or yogurt tubs)
- heavy-duty scissors
- masking tape
- pencil
- penny
- partner

you were when you were wearing your sonic ears? If not, then your sonic ears really helped your hearing. If you were, then it's back to the drawing board to design a better pair of sonic ears!

OTHER THINGS TO TRY

Try putting your sonic ears on and pressing one ear against a wall. Can you hear people talking in the other room, on the other side of the wall? If so, mind your own business! It's not nice to eavesdrop!

F·U·N F·A·C·T·s

- Auricles are rather like fingerprints—no two persons' are exactly alike. For this reason, detectives sometimes use auricles to identify people!
- Ludwig Van Beethoven was completely deaf when he wrote his ninth symphony.
- The smallest bone in your body is in your ear. It is called the *stirrup*, and without it you cannot hear.

MUSICAL INSTRUMENTS

What is your favorite kind of music? Rock? Classical? Blues? Country? Heavy metal? They may all sound different, but they have one very important aspect in common: Their sounds are all made possible because air and objects vibrate. Remember earlier in this chapter we explained that sound waves are really just the vibration of air molecules? Well, each instrument used to create music causes the air to vibrate in a different way. There are three classes of instruments, based on the method they use to create the air vibrations: *stringed* instruments, such as a harp and a guitar; *percussion* instruments, such as cymbals and a drum; and *wind* instruments, such as a flute or a trumpet.

What makes one style of music different from another is in part the way the styles combines the sounds of different instruments. Country music may be more guitar than classical, which will definitely be more violin than heavy metal, which is more drum than blues, and so on! Let's take a look at some common instruments and how they produce their unique sounds. Let's even make some instruments of our own!

GUITAR

A guitar is a stringed instrument, which means its sound is produced primarily by plucking or strumming strings. An acoustic guitar is an instrument with a long neck, a boxy hollow body with a hole in it, and six strings running its length. When you pluck a guitar's strings, the strings vibrate. These vibrations cause the air molecules around the guitar, and even inside its hollow body, to vibrate. The strings are tuned to the musical scale by varying their *tensions,* that is, by stretching some tighter than others. The sound a particular string makes is also determined by where the guitarist presses down on the string on the neck of the guitar. The farther away from the *bridge*—the part of the guitar that lifts the strings up—the lower the pitch that is produced. There are all kinds of guitars: acoustic, bass, classical, electric. They may be designed to produce different sounds, but they all make the vibrations the same way—with strings!

MATERIALS

- a "guitar" body: shoe box, tissue box, plastic box
- several rubber bands of varying widths

Remove the box top and stretch the rubber bands over the entire box. Space the rubber bands apart so that you can pluck them. You can vary the size of rubber bands you use: fat ones or skinny ones. Now, pluck them to play your guitar. Do the different widths of rubber bands make different sounds? What would happen if you designed a different guitar body? Perhaps you could put the lid on the box and cut a hole in it, then stretch the rubber bands over the hole. Does the guitar sound different?

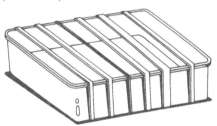

DRUM

MATERIALS

- small tub with a lid, such as a margarine or deli tub
- 15 inches of string
- beads
- tape
- wooden dowel (a chopstick or pen will work, too)
- scissors
- markers or paint (optional)

A drum is a *percussion* instrument, which means you produce sound by striking it. Other percussion instruments are cymbals, gongs, tambourines, and bells. You can make a drum out of almost anything—by drumming your fingers on a tabletop or by hitting a pot with a wooden spoon. Most drum heads, however, are made of some kind of *membrane*, like leather. The drums you are probably most familiar with are kettle drums and snare drums, which are hollow cylinders with a membrane stretched over the top; they are played with drum sticks. However, there are many different kinds of drums. The slit drum of West Africa is made by carving a slit into a log or wooden block and then hollowing out the drum through this slit. Drums are used for all kinds of purposes besides music-making. "Talking drums" are used in West Africa for communication. They may be shaped like an hourglass and can be heard more than 20 miles away! Skilled drummers can imitate the sound of their people's language, even mimicking its pitch and accents! Drums are one of humankind's oldest instruments, and were probably first used in ceremonies and sacred rituals. The drum you will make is similar to ceremonial drums found throughout the world.

(Adult Assistance Needed)
Make sure the lid is securely fastened to the container. Ask an adult to help you use the scissors to make four holes in the plastic container as follows: Poke one hole each in opposite sides of the container big enough for the dowel to slide through. Slide the dowel through the holes so that the ends of the dowel stick out each side. Now, punch two more holes in the side of the container halfway between the dowel ends, and thread the string through the holes until there are about five inches of string hanging out each hole. Tie knots in the string where the ends emerge from the container so the string will not slip back into the container. Slip beads onto the string on each side and tie them in place. These will serve as "drumsticks." You can decorate the rest of the drum any way you like.

Now, tip the container on its side so one end of

the dowel is at the "top" and the opposite end is the "bottom." Take hold of your drum by placing the bottom dowel between your palms, and then "rub" your palms back and forth to get the drum spinning. The beaded "drumsticks" will strike the drum head and make music! Alter the spinning speed and other movements of the drum to change the way it sounds.

FLUTE

A flute is an example of a *wind* instrument. Other wind instruments are a bassoon, a piccolo, a clarinet, a trombone, and a harmonica. But flutes are one of the oldest wind instruments, and most ancient civilizations had them. A flute is a long, hollow cylinder with six (sometimes more, sometimes fewer) holes along its length. A flute usually has a delicately carved mouthpiece, called a *reed,* over which the musician blows air. When a musician blows into the flute, a stream of air travels over the reed, which causes the air to vibrate very fast. The air flows through the tube of the flute and different sounds are produced depending on which holes are covered by the player's fingers and which are left open. Modern flutes are made of metal and have finely designed mouthpieces and "keys," like stoppers, over the holes that are operated by the pressure of a player's fingers. But most flutes throughout history were made of wood and were very simply designed. You can even make a flute out of a drinking straw!

ACTIVITY

ACROSS

2 The folded disk of skin and cartilage that surrounds the human ear.

5 The flute, bassoon, piccolo, clarinet, trombone, and harmonica are all examples of this type of instrument.

8 How high or low a sound is.

9 When a wave of motion brings objects closer together, it is described as _____.

11 A unit that scientists use to measure how fast or slow a sound wave vibrates.

12 Sound waves travel down your ear until they hit this tiny structure.

13 The parts of a guitar that vibrate to produce sound.

14 The units scientists use to measure the loudness of sound. These units are named after their inventor, Alexander Graham Bell.

15 The smallest bone in your body, without which you could not hear.

DOWN

1 When a wave of motion causes objects to separate.

3 When a musician blows into a clarinet, a stream of air travels over the _____, which causes the air to vibrate very fast.

4 The auricle captures sound waves and funnels them into the hole in your head called the ear _____.

5 You can change the pitch of your bottle pitch pipe by varying the amount of _____.

6 Sound moves faster through a _____ than it does through a liquid or a gas.

7 A quivering motion that usually occurs when something moves very quickly up and down or forward or backward.

8 A drum is an example of this type of instrument.

10 When a vibrating object sets the air molecules around it vibrating, the air molecules in turn set their neighbors vibrating too, which creates a growing _____ of vibrations.

MATERIALS

• plastic drinking straw
• scissors

To make the mouthpiece reed, flatten one end of the straw by biting it closed. Leave just a tiny crack of an opening. Then, carefully use the scissors to cut a tiny point into this flattened end. Hold the reed end of the flute loosely in your mouth and gently blow into the straw. Did you produce any sound? If not, keep trying. Playing the flute is fairly difficult and takes practice. Once you get the hang of it, though, you will see that your flute produces beautiful music without even having any holes! Try cutting the straws to make flutes of different lengths. How does the length of the straw affect the pitch?

CHAPTER 10

BUILDING BLOCKS

ARCHES
·············
GEODESIC DOMES
·············
TENTS
·············

INTRODUCTION

Bee hives have it; so do trees. Humans have it; so do turtle shells and sea shells. What is this mysterious "it" that all these objects and animals have? Structure. *Structure* is an organized arrangement of patterns that gives an object or animal its form. A beehive is made up of a repeating pattern of hexagonal shapes. Tree branches grow out from the tree trunk in distinctive, repeating patterns. Humans have a symmetrical body shape of two arms and two legs attached to the "trunk" of their body, and the whole body is structured by the shape of the skeleton. Many turtle shells grow in a pattern of boxlike shapes. And many sea shells form beautifully symmetrical spirals.

Although nature has designed some of the most exquisite structures, humans are very good at building structures, too. We build bridges and houses and domed football stadiums and cars and rocketships and chairs and beds and all kinds of other structures. Different kinds of structures are designed in a particular way because they have a specific function and are subject to specific forces. A *force* is anything that influences an object, but an easier way to understand force is to think of it as anything that pushes or pulls on a structure. One of the most basic and important forces is *gravity.*

Gravity is an "attractive" force, which means it is a force that pulls objects toward each other. For example, the Moon stays in orbit around the Earth, instead of flying off into space, because of gravity. Gravity is what keeps our feet on the ground, too. But at the same time that gravity is pulling us toward Earth, it is pulling the Earth toward us! Rocketships must have engines that create enough power to overcome the force of gravity.

But there are many other forces besides gravity that affect structures. For instance, besides being designed to have an engine, a car and an airplane must also be designed to move through air smoothly, that is, to overcome the force of the air rushing past them, which can slow them down or make their ride very rough. So a car's shape and an airplane's shape must be *aerodynamic:* they both have curved, smooth shapes that allow air to pass over them easily.

Sometimes, rather than having to withstand a force, a structure uses a force to its benefit. For instance, tents, bookshelves, buildings, and bridges use *balanced* forces to remain upright and intact.

To understand a balanced force, imagine a train chugging along a track. There's an engine up front that is providing the power to pull the cars of the train. The direction of motion is in the direction the engine is pulling the train. Now suppose that there is an engine at the front and at the back of the train, both trying to pull the train, only one is moving forward and the other backward. What would happen to the motion of the train? Nothing! It would not move! At least, it wouldn't if both engines had the same amount of power. That is because the pulling force would be equal in both directions, and so the train wouldn't move an inch either way! The forces acting on the train would be balanced. You could also think of two equally strong teams pulling on a rope in a tug-of-war. With a balanced force, neither team would move. Believe it or not, such balanced tugs-of-war are going on all around you in the walls of buildings, in bridges, in archways, and so on.

Let's use a bridge as an example. Bridges are fun to cross, and if they're really high they can even be thrillingly scary! Bridges are designed to

take advantage of two forces—tension and compression—working in balance. *Tension* is a force that tries to separate things, to pull them apart. When you are walking a dog on a leash and the dog tries to pull away, perhaps trying to chase after a squirrel, the leash is undergoing tension. So is your hand and arm! When things are under tension, they tend to stretch. The leather leash may stretch a little under the tension of your unruly dog, and your arm may feel like it is stretching. You could actually "pull" a muscle if your dog pulls on your arm too hard. But the best example of tension is a rubber band. Because it is made of rubber, it stretches as the tension increases, until—snap!

Compression, in contrast, is a force that pushes things together. Hold a peach too tightly and you'll compress it until juice is running down your hand! Step on an empty milk carton, and whoosh! As the milk container compresses, the air is forced out in a noisy rush.

So what does all this have to do with a bridge? Imagine a very simple bridge, perhaps a plank of wood laid across the banks of a river—no, let's make this bridge span an alligator-infested swamp!

You carefully walk across the plank of wood. As you walk out onto the bridge, the weight of your body pushes down on the wooden plank, causing its top surface to be compressed. The compression causes the top surface to shorten slightly. If the bridge is anchored on either bank of the swamp, this compression/shortening causes tension on the plank as well. This means the bottom of the plank is feeling a pulling force from the two ends of the plank, and so it is being pushed upward. This push upward lengthens the plank. So the top of the plank is pushing down and the bottom is being pulled up, which means the forces are trying to balance out. If they do balance, you are lucky—the bridge stays intact and you do not get dumped in the swamp of hungry alligators!

As you can now see, all structures are subject to forces, from the Eiffel Tower to the Empire State Building to a two-car garage to a fort made of couch pillows. And they would all be just a pile of bricks, wood, stone, metal, or pillows if it weren't for balanced forces.

ARCHES

Imagine you lived long, long ago, when humans decided it was time to give up living in caves and to make proper homes. What kind of house do you suppose they designed? No doubt it was very simple, perhaps just a domed hut or a square box with a crude opening to get in and out. Before too long, however, humans discovered how to make their homes more comfortable and functional. They added things like doors and windows. That was a big step for those early architects, because doors and windows are not so easy to build. Now you are probably wondering, what is so difficult about building a window into a house? But stop and think for a moment. If you cut a hole in a wall for a window, what would support the wall above the opening? And if you were building a mud-brick home, how would you fill in the wall above the window opening? How would the bricks stay up? We think you will agree that windows presented problems for early home builders. Until, that is, they discovered the arch! Then all their problems were solved.

An *arch* is a curved structure that spans an opening and serves as a support. You may have seen arches in the interior of a church, library, or courthouse. Because arches can support a lot of weight, they are often used to support the roofs of large buildings. They are more commonly used in the design of doors and windows, because they not only support the weight of the wall, but their shape leaves a natural opening. The first arches were very simple. They were made from little more than two columns with a beam laid across the top. A *column* is a vertical (upright) support. Columns are very strong and can withstand a lot of compression force, which is a fancy way of saying that they can support a lot of weight. You can find columns all around you. Tree trunks are natural columns, as are dandelion stems; your legs are columns, too! So are telephone poles and fence posts. Early columns were made from stone. To make the archway, the ancient architect would place two stone columns a certain distance apart and then lay a stone beam across the tops. The beam is called a *lintel,* which is a horizontal support that can also carry a lot of weight. Look at the door or window closest to you: Its top frame is probably a lintel.

Of course, as the early builders became better architects and began building bigger homes and huge churches and public buildings, they did not want to settle for boring square windows. Or small ones. Columns and beams could not span very great distances before they broke under their own weight. So these builders were forced to design a better arch. Which they did! It is called a *cantilever arch,* and it looks something like two upside-down staircases that lean on one another. That may sound odd, but it works! You see, each "step," or cantilever, is really just a beam that is supported on one end. Think of a diving board supported at one end at the edge of a pool. Now stack another diving board on top of it but push it back a little, so the front edges of the diving boards are staggered. Eventually you will get a staircase effect. Each cantilever rests on the one below it, pressing down or compressing it. Because of all this compression, no single cantilever has to bear all the weight, but the entire arch can bear a tremendous amount of weight because it spans a larger distance. Build one for yourself to see!

MATERIALS

• Objects that are "block-like": wooden blocks, books of equal size, bricks, decks of cards.

I-N-S-T-R-U-C-T-I-O-N-S

Place one block on the floor or a tabletop, then place another on top of it but push it to the right a little (so the edges are not aligned). Now place another block on top of this one, again pushing it to the right a little. Repeat this procedure until you create a "staircase" of blocks several blocks high. Now do the same thing with another stack of blocks, only stagger them to the left, so the left edges are not aligned. Build this staircase to the right of the first one. When you have an equal number of blocks stacked in this staircase, push the two staircases together to form a cantilever arch.

OTHER THINGS TO TRY

Once you become an expert cantilever-arch builder, try to see what is the longest distance you can span using just ten blocks to create an arch. What is the tallest cantilever arch you can build?

GEODESIC DOMES

Cut a hollow rubber ball in half and what do you get? You get two domes. A *dome* is a hemispherical shape like the half of a hollow rubber ball. A turtle shell is a dome; so is our skull. An egg is two domes put together. Domes are everywhere in nature because they are easily formed but very strong structures. That is why humans have copied nature and used the dome shape in many buildings, such as atop the Capitol building in Washington, D.C., and as the design for many sports stadiums and for the main building at Disney's Epcot Center. In Antarctica, scientists constructed a huge dome-shaped building that is 50 feet high and 160 feet in diameter! A dome is a good design for their research building because it can be built big and from very lightweight materials, yet the dome shape makes it very strong, so it can withstand the force of the strong winds in Antarctica.

Most dome-shaped buildings are called *geodesic domes*, because their design is really a pattern of interlocking triangles. The geodesic dome was invented by a man named Robert Buckminster Fuller, who first became interested in the triangle shape when he was in kindergarten! His teacher gave everyone in the class a handful of toothpicks and peas and asked them to create structures. Bucky Fuller stuck two toothpicks into a pea at an angle and found that the triangular shape he had created did not roll over very easily. It stayed in place, or was *stable*. Later, when he grew up, he remembered this fact from his kindergarten days, and that knowledge helped him create a building whose design was based on a series of interlocking triangles—the geodesic dome!

So just why is a triangle so strong? The secret is in how its shape withstands the forces that act upon its three sides. When a triangle is part of a structure, two of its sides usually are being compressed, or pushed upon. The other side, called the *base*, is under tension, or being pulled. The balance of the compression and tension forces keeps the triangle from being flattened out. Sound complicated? Well, think of it this way. Imagine a triangle with a point at the top; then remove the base side, so the shape now looks like a tent or an upside-down "V." If you pushed down on the point of the triangle—that is, if you compressed it—the two "legs" would eventually collapse under the compression force. But add the base side back in and this problem is solved. Now, if you push down on the point, this base "leg" pushes back up in a tension force, pulling the two legs toward it and keeping them from flattening out. Thus, it is very difficult to flatten the triangle. That is why triangles are strong. And that is why a geodesic structure is very strong as well. To explore the properties of this shape yourself, you can easily build a geodesic dome.

MATERIALS

- 25 toothpicks
- 11 gumdrops or small marsh-mallows

Connect five toothpicks and five gumdrops together to form a *pentagon* (a five-sided shape).

Now insert two more toothpicks into each gumdrop. Using five more gumdrops, connect each new pair of toothpicks to form five "V" shapes.

Your original pentagon should now have five upright triangles sticking out of it. Now take five toothpicks and insert them horizontally to connect the five top gumdrops together. Your dome should now have taken form and should be ready to have its top constructed. To construct this top, insert five toothpicks around the outside edge (the circumference) of a

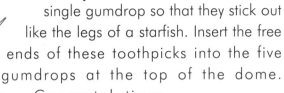

single gumdrop so that they stick out like the legs of a starfish. Insert the free ends of these toothpicks into the five gumdrops at the top of the dome. Congratulations, you have just built a geodesic dome!

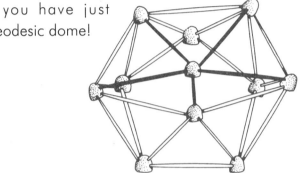

F·U·N F·A·C·T·S

Buckminster Fuller said, "There are no failures—only unexpected results." What do you think he meant by this?

TENTS

If you have ever been camping, chances are you have slept in a tent. Tents are lightweight, portable structures that can protect you from rain, cold, wind, and blood-thirsty mosquitoes! Traditionally, tents were made of heavy canvas and were shaped like an upside-down "V." Today, however, tents are high-tech, made of space-age plastic materials that are lightweight and easy to care for. Modern tents come in all sizes and shapes. They are used for camping, as emergency housing after a disaster, as sun shields at the beach or a backyard wedding, and, if they are really humongous, to house a circus! Native American nations such as the Lakota, Blackfoot, Cheyenne, and Crow were supreme tent builders. Their tepees were conical-shaped tents—like upside-down ice cream cones—made of animal hides draped over wooden poles.

Whether a tepee, a one-person "pup" tent, or a giant circus tent large enough to shelter six elephants and everyone in a small town, tents are useful structures that are relatively easy to transport and assemble, yet are strong and protective. They have these qualities because their shapes are extremely good at creating balanced forces. Remember that we said tension is a pulling force and compression is a pushing force. Well, in a tent these two forces are balanced and cancel each other out, making tents very strong structures. The strings used to stake a tent to the ground are under tension. They are pulled very tight. The center pole of the tent is under compression because it is pushed down by the tent strings. It is pushing back up against the strings. If the strings are pulling with the same amount of force as the pole is pushing, the forces balance and the tent is very stable: It does not move easily and can withstand a blast of strong wind. If the forces become unbalanced, perhaps because the stakes loosen and the strings begin to sag—well, then, watch out! Your tent may be flapping in the wind!

MATERIALS

- 1 drinking straw
- 4 thumb tacks
- 1 piece of cardboard
- 2 pieces of string, each 2 feet long
- paper
- glue
- scissors

Cut four small notches in one end of the drinking straw. The straw will be positioned in the middle of the cardboard as the tent pole after the strings have been attached. To attach the strings, tack one end of a piece of string to one corner of the cardboard. Then run the string up over one end of the straw, anchoring it in two notches, and across the cardboard to the other opposite corner. Tack it in place there. Tack the other piece of string to

ACROSS

1 An organized arrangement that gives an object or animal its form.
3 He invented the geodesic dome.
5 Before humans learned to build their own homes, they lived in these natural structures made of rock.
8 A car's shape and an airplane's shape must be _____ to allow air to pass over them easily.
12 A horizontal support that can carry a lot of weight.
14 A beehive is made up of a repeating pattern of these types of shapes.

DOWN

2 A type of arch made of steps.
4 A force that tries to separate things.
5 A force that pushes things together.
6 In order for a bridge to support your weight, the forces of tension and compression must be

_____.

7 An attractive force that pulls objects toward each other.
9 A hemispherical shape like the half of a hollow rubber ball.
10 A vertical support.
11 A type of dome with a design based on interlocking triangles.
13 A lightweight, portable structure that people use for camping.
15 A curved structure that spans an opening and that serves as a support.

a free corner of the cardboard, thread it over the top of the straw in the same way, and tack it to the opposite corner of the cardboard. If you need to, position the straw upright in the middle of the cardboard platform, forming a tent structure. Cover the structure with paper, and cut a flap for a door.

Can you see how tightening the strings pulls down on the straw?

F·U·N F·A·C·T·S

Here's a joke to tell your friends. Why are tents called "tents?" Because they are under a lot of stress! (If your friend doesn't get the joke, explain the word play of "tents" and "tense!")

CHAPTER 11

WATER, WATER EVERYWHERE

INTRODUCTION

What are you made of? Flesh and bones and blood and brain and muscle and skin, right? Right! Your body is made of all these things. But did you know that you are mostly water? You are literally a body of water—65 percent water! Actually, all living things are mostly water. A tomato, for example, is 95 percent water; a watermelon, 92 percent water. A potato is 80 percent water, and an egg, 74 percent. Even seemingly dry things like grass and birds are mostly water. Now, these may be pretty interesting facts, but water itself may not seem so interesting. After all, it is colorless and odorless and tasteless. Besides being essential to life, you probably think there's nothing else interesting about water. Right? Wrong!

Water is everywhere, probably because it is made up of two very common elements, hydrogen and oxygen. When two hydrogen atoms link up with one oxygen atom, they make water. The chemical formula for water—yes, water is a chemical!—is H_2O. This formula tells us that two hydrogen atoms link up with one oxygen atom. If you could see a water molecule, you might think it looked a little like Mickey Mouse's head—Mickey's face would be the one large oxygen atom, and his ears would be the two hydrogen atoms.

Like a magnet, each water molecule has two "charges"—one end is positive and the other end is negative. You know the saying "opposites attract?" Well, the positive end of one water molecule attracts the negative end of another water molecule. Why are we telling you all this? Because these charges explain why water molecules love to come together in groups—to form puddles and pools and lakes and oceans! Mickey Mouse's hydrogen-atom ears are positively charged, but his oxygen-atom face is negatively charged. Therefore, the "face" of one water molecule is attracted to an "ear" of another H_2O molecule, and they link up. This happens over and over again, until there are millions of water molecules all linked together. Water molecules simply love to stick to each other. Don't believe us? Get an eyedropper and see how many water molecules you can pile on the flat side of a penny!

You drink water, you bathe in water, you are mostly water, but you would probably deny that you can stand on water. But you are wrong! You can stand on water if it is frozen! Ice, after all, is water. Water exists in three different states: liquid, solid, gas. You drink liquid water and suck on solid water (ice cubes, Popsicles); you swim in liquid water and skate on solid water. Water turns from a liquid state into a solid state at 32 degrees Fahrenheit (zero degrees Celsius), its freezing point. Why does it freeze? Because as water molecules lose heat, their motion slows down and they become very rigid. Remember how we said that water molecules love to stick together? Well, in their liquid state they stick together loosely, and they can move around fairly easily. That's why liquid water runs and pours and flows. But as water loses heat, the molecules slow down and stick together more closely. They become almost rigid, forming ice.

At 212 degrees Fahrenheit (100 degrees Celsius), water boils, moving from a liquid state into a gaseous state. Gas? That's right. As water molecules heat up, they move faster and faster—so fast that some of the molecules begin to release their hold on their neighbors and drift off into space as individual water molecules. Fog and steam

are examples of water in a gaseous state, although we usually call it "vapor." On humid days, or if you are in a tropical jungle, water vapor literally hangs in the air. You may not see it, but you can certainly feel it. If you breathe onto a mirror, you can see water vapor escaping from your body through your breath.

Clouds are little more than huge masses of water vapor. When this water vapor sticks together, it forms little droplets you know as raindrops, which fall to the ground. Then the heat of the sun "evaporates" the water on the ground, pulling the water vapor upward to the sky again. If the air temperature is cold enough, water may fall from the sky not as a liquid (rain) but as a solid (hail)! So you see, water is continually recycled. Naturally!

ICE CREAM

Ice is pretty cool! And it can be a lot of fun. Just think of icicles, and ice skating, and ice cream! You already know that ice is frozen water. As we explained in the chapter introduction, ice is water that has lost a lot of its heat. Water molecules, when in liquid form, can move around a lot; they can hitch and unhitch themselves from one another fairly easily. But as the molecules lose heat, their movement slows down and they lose their ability to unhitch themselves from their neighbors. That doesn't mean they bunch up in tight packs, but only that they form loose bonds and hold that pattern very stubbornly in what is called a *lattice*. Although ice is more rigid than water, it is less *dense*. Its molecules form looser bonds than liquid water molecules—there is more space between them. So ice will float on water. Even though ice feels heavier than water, it is less dense, so it floats!

People who live in cold climates are familiar with ice floating on the surface of a bowl of water. The bowl of water is called a pond! In winter, ice forms on the surface of ponds and small lakes. People can have fun ice fishing and skating. During cold weather, ice will form anywhere the air temperature falls to 32 degrees Fahrenheit (zero degrees Centigrade) or lower. Ice is fun when it forms on a lake or a skating rink, but it is dangerous when it forms on walkways or sidewalks. Fortunately, chemistry can come to the rescue—with salt! Remember we said water molecules can switch back and forth between solid and liquid (and gas, too)? Well, as the ice begins to melt into liquid, the salt prevents that liquid from turning back into a solid. So once the ice on your walkway begins to liquefy, the salt keeps the water molecules liquid and you can walk safely. In order for salt to do its job, it needs heat, and it gets heat from the environment—the surrounding air and ground. When you make ice cream, the salt and ice take heat away from the cream and sugar. So much heat is removed that they turn into ice cream.

MATERIALS

- 2 sandwich-size Ziplock baggies
- 2 cups of ice
- 8 tablespoons of salt (kosher salt works best)
- ½ cup of milk
- 1 tablespoon sugar
- ¼ tablespoon vanilla

MILK BAG

In one bag, add the milk, sugar, and vanilla. Squeeze out as much air as possible, then seal it closed.

ICE BAG

In the other bag, put 1 cup of ice and 4 tablespoons of salt. Place the sealed milk bag inside the ice bag, on top of the ice. Then, pile the last cup of ice and four remaining tablespoons of salt on top of everything. Seal the bag.

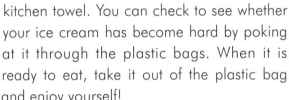

SHAKE, SHAKE, SHAKE

Now, shake the bag around for about ten minutes. It may get pretty cold—remember that the salt is making the ice colder. So you may want to hold it wrapped in a kitchen towel. You can check to see whether your ice cream has become hard by poking at it through the plastic bags. When it is ready to eat, take it out of the plastic bag and enjoy yourself!

Make sorbet by using fruit juices instead of milk.

Experiment with different flavors; try adding chocolate or strawberry syrup or crushed cookies.

What happens if you try making ice cream without using salt?

FUN FACTS

• Popsicles were originally called "epsicles," because the man who invented them was named Mr. Eppson. He didn't actually invent them; he discovered them by accident one winter when he found a frozen glass of lemonade with a spoon in it!

• In the United States, an average person eats 14 quarts of ice cream each year.

BUBBLES

Just about every kid (and an adult or two!) has blown bubble gum bubbles. If they get too big, you end up with gum all over your nose! You've also probably lathered your soapy hands into a bubbly froth or blown air through a straw into your chocolate shake to make it boil with thick milkshake bubbles. And you've no doubt blown soap bubbles of all shapes and sizes—some as small as a penny and others as big as a baseball glove! Soap bubbles are particularly fun because you can blow really weird bubbles—double bubbles and tubular bubbles that look like translucent worms!

Bubbles are very simple to make—all you need is soap and water. Soap bubbles are mostly water, and water has a unique quality that makes it an excellent bubble ingredient—it loves itself! Well, not really. Water does not have emotions. What we mean is that water molecules are strongly attracted to each other; they love to stick together. As you remember from the beginning of the chapter, a positive end of one water molecule attaches to the negative end of another water molecule. That is why a drop of water has a round shape: the water molecules want to get as close together as possible, so they all "hug" each other. They are hugged left and right and up and down. The water molecules on the outside of the drop are also being tugged from their bottom by all the water molecules inside the drop. This tugging causes a kind of "skin" to form on the surface of the water droplet. Now, this isn't a skin like the skin of a grape. It cannot be peeled off. It just acts like skin—encasing the water molecules so they stay inside the droplet. Of course, if you put too much pressure on this skin, it bursts, and the water droplet loses its shape.

You can see this water "skin" by filling a glass of water to the tippety-top. Then use an eye dropper to carefully drop water droplets one by one into the glass. The water will mound up in the glass, droplet sticking to droplet, because of this sticky skin. This mountain of water is called a *meniscus*. In the end you will add so much water that the stickiness of the skin gives way and water overflows the glass. But until that happens, the meniscus will keep building, one water molecule piling up and hanging on for dear life to another!

It is hard to blow bubbles in plain water precisely because this water skin is so sticky. The water molecules hug each other tightly and resist being pulled apart or puffed up into a bubble by the air you are blowing into the water. But add a drop of soap and your problem is solved—bubbles! Soap loosens the bonds of the water molecules so they will stretch a lot more easily. Then when you blow air into the water, the skin of each water molecule—having been relaxed by the soap—expands. If you are lucky and your bubble solution is well mixed, it is possible you could blow a bubble more than 30 feet long!

MATERIALS

- clean, cold water
- liquid dishwashing soap (Dawn and Joy work really well)
- clean dish pan or tub
- glycerin (optional; available at most drugstores; it makes bubbles even stretchier
- 1 or more pipe cleaners
- string (about 3 feet)
- 1 or more paper cups

INSTRUCTIONS

Add one cup of the dishwashing liquid to about ten cups of water in the tub or pan and gently stir them together. Stir in four tablespoons of glycerin for extra-stretchy bubbles (it helps you blow them really big!). Blow bubbles using the bubble wands described below.

Bubble Wands

Paper Cup Wand: Poke a hole in the bottom of a paper cup, dip the rim of the cup in the bubble solution, then gently blow through the hole into the cup to make bubbles.

Pipe Cleaner Wand: Bend one end of the pipe cleaner into a shape and dip that end into the bubble solution. Then, gently blow to make a bubble. Try a circle, a square, a triangle, and a star. What shapes work best? You can also combine pipe cleaners to make three-dimensional bubble wands. Try making a cube or pyramid shape and see what happens when you dip the whole structure in the tub of soap solution.

Straw String: You can create giant-size bubbles using this device. Cut a three-foot-long piece of string. Thread the string through the two straws and tie the ends together, making a loop. Now, hold one piece of straw in each hand, so the string is hanging between them, and dip the whole device— your hands included—into the bubble solution. Remove your hands from the solution and slowly draw out the straw string device, pulling your hands apart so the string stretches out between them and separates into a loop. You should see a sheet of soap solution clinging to the loop of string. Now gently blow into the loop or drag the loop through the air, creating a giant bubble!

NOTE: Bubbles work best on very humid days.

CLEANING TIP
Bubble solution is not always easy to clean up, because the

more water you add, the more bubbles you get! Just add a bit of vinegar to the solution, and all those bubbles will disappear.

STORAGE

Store your bubble solution in an airtight container. It keeps for a long time, and actually gets better with age!

F·U·N F·A·C·T·S

- The writer Mark Twain said, "A soap bubble is the most beautiful thing, and the most exquisite in nature. . . . I wonder how much it would take to buy a soap bubble if there were only one in the world?"

- Eiffel Plasterer, a retired high school physics teacher and king of making long-lasting bubbles, is said to have kept a bubble in a jar for 340 days!

CHROMATOGRAPHY

A *solvent* is a liquid capable of dissolving another substance. Water is a universal solvent—that means it will dissolve just about anything in the universe. Of course, there are some substances that do not dissolve well in water: oil and certain plastics, for example. But most other substances and objects, from a powdered fruit drink to a Jeep, will dissolve if you leave them submerged for long enough. The powdered fruit drink will dissolve in seconds, but the Jeep may take thousands of years!

Water usually contains tiny pieces of the stuff that has dissolved in it. Ocean water, for example, has molecules of salt, oxygen, and minerals in it, as well as tiny particles of plants and dead fish and many other things. Water that has something dissolved in it is called a *solution*. The stuff that is dissolved in water is called the *solute*.

One way to separate water from the solute is to filter the water. For example, if you pour ocean water through a coffee filter, any solute larger than the holes in the coffee filter will remain on top of the filter. Another way to separate solute from a solution is to boil it. If the solvent were water, the steam would be pure water, and the solute would be left sitting in the bottom of the pot. Much the same process would occur if you left the solution out in the sun. The heat from the sun would cause the water to evaporate into the air, and it would come down to Earth again someday as rain! This is the natural process by which we get clean drinking water. The Earth works as a filter: The water is evaporated up as pure water, and the solute (anything in the water, from oil pollution to dead leaves) is left behind in the soil.

Because water can dissolve and filter substances so well, scientists have found ways to use this principle to help them study nature and to invent technologies. For instance, a technique called *paper chromatography* was developed to separate the solute from a solution. In paper chromatography, the solute is something that a scientist wants to separate into tinier parts— something like black ink. The powerful dissolving properties of water are used to separate the solute. Now, you might think that black ink is pretty simple, and that it would be hard to separate it into tinier parts, but hold onto your hats because black ink is actually made by mixing a lot of colors together. You can separate black ink and see these colors by doing a paper chromatography experiment.

As we said before, a solution has something dissolved in it, such as powdered fruit drink. But when doing paper chromatography, the solution is not directly mixed in with the water to begin with. First, a line of black ink is drawn onto a piece of paper (hence the name "paper chromatography"). Then just a little bit of the bottom edge of the paper is placed in some water. As the water begins to soak the paper, it travels up toward the ink line. When the water and the ink meet, they become a solution. The water continues traveling up the paper, taking parts of the ink with it. Some of the colors that are mixed to make black ink have heavy molecules, and these colors get left behind as the water travels up the paper. The colors that have lighter molecules follow the water up the paper toward the top. The piece of paper ends up looking like a rainbow! All this beauty from a line of black ink, a piece of paper, and some water!

Not all black inks are made from the same mixture of colors. The FBI, for instance, uses paper chromatography to identify what type of pen may have been used to write a ransom note or a bad check. It's time to try this technique yourself. How good a detective are you?

MATERIALS

- paper coffee filter or heavyweight paper towel
- black felt-tipped marker (nonpermanent)
- water
- clear drinking glass or cup
- tape
- pencil

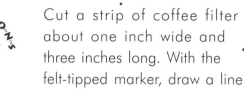

Cut a strip of coffee filter about one inch wide and three inches long. With the felt-tipped marker, draw a line across the middle of the filter strip. Tape one end of the filter strip to the pencil. Now, fill the glass with an inch ot two of water and place the pencil across the top of the glass, so the filter strip hangs down into the glass and is just barely touching the water. Be patient! It will take three or four minutes for the paper to absorb the water. Don't rush things and move the filter or the glass. What happens to the ink line on the filter strip as the water travels up the filter?

CRIME SOLVING

The ink in most black felt-tipped markers is made of several different pigments dissolved in water; these appear black when mixed together. Not all kinds of black felt-tipped markers have the same type of ink in them. You can use paper chromatography to separate the various colors in different types of black marker ink.

When you separate the inks using paper chromatography, you can distinguish these different inks because each leaves a different pattern—a kind of "fingerprint." Provide a family member or a friend with several different makes of black felt-tipped markers and ask him or her to write a "ransom note" on a piece of black coffee-filter paper. Then perform chromatography tests on the ink from each pen and the ink from the ransom note. Can you identify which pen the ransom note was written with?

How do you think the ink from different colored markers will separate? Which color ink separates most easily? What happens when you test the ink from a permanent marker?

Oil and water do not mix. Oil will never dissolve in water.

FLOATING AND SINKING

Drop a rock into water and it quickly sinks out of sight. However, put a boat—which is much heavier than a rock—into the water, and it floats. Of course, you already know that light objects like Styrofoam float on water. Have you ever wondered why other things float or sink? The answer has to do with density, uplift, and displacement.

Density is the relationship between an object's volume and its mass. *Volume* is the amount of space the object occupies, and *mass* is a measurement of the amount of stuff in the object, sort of like weight. An object with a lot of mass can have a low density if it is very big, and an object with little mass can have a high density if it is very small. So you can increase an object's density either by condensing it to fit into a smaller volume or by adding more mass to it. Similarly, you can decrease an object's density either by expanding it to fit into a larger volume or by taking mass away from it. Consider three balls of the same size but made from different materials. A ball made of Styrofoam has less density than one made of rubber, which in turn is less dense than a lead ball. This is because their masses are different, even though their volumes are identical.

But floating and sinking involve more than density—water also plays an important role. Water has an interesting property called *uplift*, which means that water pushes up against objects that are placed on its surface. Of course, good old gravity tries to pull the object down through the water, just as uplift tries to push the object back up. If you place a Styrofoam ball in water, the water pushes up against it, and the Styrofoam ball floats easily. The Styrofoam is less dense than the water that it tries to sink through, so it floats. Therefore, we describe the Styrofoam as less dense than the water. However, drop a ball of lead into the water, and it sinks straight to the bottom. This is because the mass of the lead is greater than the mass of the water that the ball tries to sink through. Therefore, we describe the lead as more dense than the water. The lead is so dense that its downward thrust exceeds the water's uplift capacity, and it sinks. Now you know why life savers are made from Styrofoam, not lead.

But what about ships? Ships are made of materials that are denser than water, yet they float. This is where displacement comes in. Displacement is related to uplift. *Displacement* is what happens when an object pushes water out of its way. Water doesn't like being pushed around, so it pushes back against the object, trying to return to the space from which it was pushed. Thus, the effect of the water pushing against the ship makes the ship float. Try it for yourself!

MATERIALS

- tub of water
- modeling clay
- rocks of various sizes
- various fruits and vegetables

Make a ball about the size of a Ping-Pong ball out of the clay and place it in the water. It will sink, because it is heavier than the small amount of water it displaced. Now make a flat-bottomed boat out of clay. Put it in the water. What happens? It floats! That's because the shape of the clay and the air inside the boat will keep it afloat. Try experimenting with different shapes to see which float and which do not.

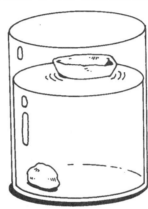

ACROSS

1 When water gets hot enough, it turns into

_____ .

2 Water that has something dissolved in it.
3 When an object pushes water out of its way, it causes _____ .
6 This substance can keep water from freezing, even when the temperature is below 32 degrees Fahrenheit.
7 A liquid capable of dissolving another substance.
11 The surface of a water droplet acts like a

_____ .

12 When a glass of water is completely full, the surface of the water forms a curved mound or _____ that rises above the rim of the glass.
14 The amount of space an object occupies.
15 Water in its solid form.

DOWN

1 The stuff that is dissolved in water.
3 Ice floats on water because it is less _____ than water.
4 The two hydrogen atoms in a water molecule have _____ electrical charges.
5 A measurement of the amount of stuff in an object.
8 Clouds and fog are made of this form of water in its gaseous state.
9 The oxygen atom in a water molecule has this type of charge.
10 The relationship between an object's volume and its mass.
13 Water pushes up against objects that are place on its surface. This is called _____ .

CHAPTER 12

ANIMAL TRACKS

INTRODUCTION

You animal! You *are* an animal, you know. Just as a pig, a buffalo, and a worm are all animals. Animals, as you can see, come in all shapes and sizes. They can look very different from each other. For example, gorillas, chimpanzees, and humans look a lot alike. But mosquitoes and whales look very different. Although there are more than one million different kinds of animals (how many can you name?), all animals have two things in common: They are all at least a little bigger than bacteria, and they can move around. (Plants are at least a little bigger than bacteria, but they can't move from place to place on their own. So they're not animals!)

Because there are so many different kinds of animals, scientists have had to figure out a way to organize them for study. So they've developed a system that divides animals into two groups: those that have a backbone, and those that don't. Humans, dogs, and alligators, for example, all have backbones (you can feel yours if you reach around and run your fingers down the middle of your back). These animals are called *vertebrates.* We vertebrates also have an entire skeleton of bones,

which helps give us structure. Imagine what you'd look like without bones!

Believe it or not, 95 percent of the animals in the world do not have backbones. These animals are called *invertebrates.* The invertebrate group includes animals such as insects, spiders, worms, lobsters, snails, clams, squids, and jellyfish. Some invertebrates have an outer shell that gives them structure and protects them from harm. But other invertebrates have no shells. They usually live in the water, so they don't need the same kind of support in order to move around as do the invertebrates that live on land. These invertebrates have evolved clever ways to protect themselves. For example, sea anemones make poison and squids squirt black ink.

Even though scientists who study animals—they are called *zoologists*—make their jobs easier by grouping animals into vertebrates and invertebrates, they still face a lot of challenges. For instance, it can be really confusing trying to name all the different kinds of animals that exist. One problem is that people speak so many different languages. If you tell someone who speaks only Spanish that a "bear" is charging at her, she might

not run, because the Spanish word for bear is *oso!*
In addition, an animal might have several names.
For example, in English, the red wiggler worm is
also known as the red worm, manure worm, red
hybrid, fish worm, dung worm, and at least half a
dozen other names! To solve these problems,
scientists have developed a way to name animals
(and plants and other living things) so that
scientists speaking different languages will still
know exactly what animal they're talking about.
They have given every animal a "scientific name,"
most of which are in Latin. What's more, the Latin
name usually describes the animal. For example,
the scientific name for the domestic ferret is
Mustela puturios furro. That name translates to
"those who carry off mice and stink." Ferrets do
both of those things! Bats belong to the *Chiroptera*
family. Chiroptera means "hand wing," because a
bat's wing is really its hand. *Tyrannosaurus rex*
means "tyrant lizard king." You are a *Homo
sapiens*, a "thinking man." By using an animal's
scientific name, scientists all over the world can
understand each other.

BIRD FEEDER

"Look up in the sky! It's a bird, it's a plane, it's Superman! No, wait! It's not Superman at all. It *is* a bird!" If you looked up in the sky right now, there's a good chance you would see a bird. Birds live just about everywhere, and there are thousands of different kinds. The smallest bird in the world is a hummingbird the size of a bee *(Mellisuga helenae)* that lives in Cuba. It is only 2½ inches long from its beak to the tip of its tail! Then there is Larry Bird—about six feet, nine inches tall. Just seeing if you are paying attention! Larry Bird is not a bird at all. Actually, Larry Bird is a human being (Homo sapiens), but there are birds who are taller than he is! Some ostriches can grow to be eight feet tall and weigh up to 350 pounds! But you won't see these gigantic birds flying overhead or perched in a tree, because ostriches can't fly! No, if you looked up in the sky or a tree, you'd probably see one of about five thousand different kinds of "perching" birds, called *passeriforms*. They are birds such as blue jays, chickadees, pigeons, owls, and eagles, whose feet are well designed for grabbing branches, which is why they are called perching birds!

Birds are so interesting, and there's so much to know about them, that this entire book could be filled with bird information and we still wouldn't be able to tell you everything about birds! However, you can begin to learn a lot about birds by watching them in your yard. Invite them over for some lunch by building a bird feeder!

MATERIALS

- milk carton (paper or plastic)
- scissors
- ¼-inch dowel or straight twig about 10 inches long
- string
- birdseed

I-N-S-T-R-U-C-T-I-O-N-S

Rinse the milk carton really well, because birds don't like milk! (Only we mammals—animals that are covered with hair and fur—like milk.) Then, cut out a fairly large window on one side of the milk carton. Don't cut the window all the way down to the bottom of the carton, because you want to have space in the bottom to hold the seed. It's a good idea to have the bottom of the window at least three inches up from the bottom of the carton. Poke a hole right under the window, and poke another hole on the other side of the carton exactly opposite the first hole. Stick the dowel

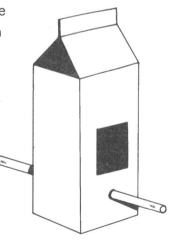

or twig through both holes so it is sticking out both sides. This is a perch for birds! Finally, fill the carton with birdseed, cap the top, and hang the feeder outside. (Birds feel most comfortable in an area where there are a lot of branches. So try to hang the feeder where there are a lot of trees or bushes.) Lunch is served!

Keep a bird journal. You can record how many birds come to your feeder, and how many different kinds. You can sketch pictures of them and write down their identifying characteristics. Your library has field guides to birds, which will help you identify birds and give you their scientific names.

Birds need water just as much as they need food. Why not build them a birdbath? To do this, set out a shallow dish such as an old casserole dish (you can find a lot of ugly old dishes at yard sales—the birds don't mind), and keep it full of water.

HERE'S ANOTHER WAY YOU CAN FEED THE BIRDS:
Smother pine cones in peanut butter and roll
them in bird seed. Hang them from branches
in the same area that you hung the bird
feeder.

FUN FACTS

- One type of ostrich egg weighs three and a half
  pounds—large enough to make an omelet for twelve
  people.

- Many adult birds regurgitate (throw up) food for
  their babies to eat.
- A brown pelican can hold up to three gallons of
  water in its stomach.

CAMOUFLAGE

Do you ever get the feeling that you are being watched by something you can't see? It's a creepy feeling, but depending on where you are, it may be true! If you are walking in the woods or through a meadow, for instance, there may be a lot of critters watching you who do not want you to see them. Creatures who don't like to be seen are usually very good at blending in with their environment. This ability is called *camouflaging*. Take lions, for example. They're good at camouflage. Their coats are yellow, just like the tall grasses in which they hide. Their coloration helps them to blend into the grass, so that they can get really close to their prey before they jump on them and eat them for dinner! But most of the animals who use camouflage are very small, such as spiders and insects, and they hide because they do not want to be eaten.

Let's look at insects and spiders more closely for a second. (They may be looking at us right now, so we might as well look at them!) To be an insect, you have to follow the insect rule: All insects have six legs and three body parts. If a critter has a different body structure, it doesn't get to join the insect club. Spiders are not insects because they have eight legs and only two body parts. Both insects and spiders belong to a large group of animals called *arthropods*. In English, *arthropod* means "jointed legs." If you look closely at an arthropod's legs, you will see that they are jointed. In some arthropods, such as crabs and lobsters, it's easy to see these joints. In others, it's very difficult. Arthropods also have their skeletons on the outside of their bodies! This skeleton isn't made of bones like ours; instead, it is more like a shell. This outside skeleton is called an *exoskeleton*. ("Exo" means outside.) The exoskeleton gives structure to arthropod's bodies and protects them.

However, as any arthropod will tell you, if something BIG wants to eat you, no exoskeleton is going to stand in its way. So, many arthropods depend on camouflage to help hide them from predators. As you know, a lot of insects and spiders hang out on plants; some have evolved to look like parts of a plant. Some look just like a tree branch, others camouflage as leaves, and still others as thorns! There is one type of spider called a "bird dropping spider." Guess what it looks like! This is a very clever camouflage, because not many predators feast on bird droppings!

If you were to design camouflage for an insect or a spider, what would you have to consider? The arthropod's size? Its habitat? Its potential predators? Try your hand at designing camouflaged critters by building a camouflage collage!

MATERIALS

- drawing paper
- coloring tools (crayons, markers, or paints)
- craft glue
- cardboard
- scissors
- recycled trash (egg cartons, aluminum foil, yarn, beads) (optional)
- twigs, leaves, grasses, and other natural items

INSTRUCTIONS

Go outside and collect some twigs, leaves, grasses, mosses, flowers, bark, and whatever else you find that might be suitable for your arthropod environment. Be kind to nature—try to select items that have fallen on the ground. Glue these items onto the cardboard, covering the entire surface. Now, create several arthropods with the paper and recycled trash. These can be imitations of real insects and spiders, or ones that you invent based on how you answered the earlier camouflage questions. Use color, pattern, and texture to create arthropods that will blend in perfectly with your arthropod environment. Glue them to the environment. Hide as many insects and spiders as you can. Ask some friends to try to find as many of them as they can. Did they miss any?

WORM BINS

What has five hearts but no eyes, makes slime, and has a brain the size of a grain of sand? No, it's not some science fiction monster. It's the earthworm—that backyard creature that seems so boring but is actually quite amazing. You are probably most familiar with the common earthworm, whose scientific name is *Lumbricus terrestris*. You may have seen them on the sidewalk after it rained, or while you were digging in dirt. But you'd be mistaken if you think all earthworms look like these. There are more than forty thousand kinds of worms, and almost three thousand types of earthworms! For example, in Australia, you'll find *Magoscolides australis*, which can grow to be 10 feet long! And in Great Britain, look for *Allolobophora cholorotica*, which is green!

An earthworm's body is actually a fairly simple structure. On the outside, it is divided into about one hundred visible segments. On the inside, a worm is mostly stomach. Earthworms eat *organic* matter—decaying bits of plants scattered on the ground. This plant matter is digested in the worm's stomach and is then excreted as dirt! That's right, dirt! In scientific language, what worms excrete is called *castings*. And castings are a major component of dirt. We can thank earthworms for the rich soil that covers the earth—that goes into gardens, baseball fields, and mud pies.

One of the best ways to get to know earthworms is to study them in action in their environment. They live underground, so this is pretty hard to do. But you can easily create an artificial environment-a worm bin!

(Adult Assistance Needed)

THE BIN: Any clean, opaque (not see-through) plastic container will make a good home for worms. You can find inexpensive plastic containers with lids at most department stores. A good size to start with is something the size of a small toaster oven. Earthworms need air, so get an adult to help you poke holes in the container.

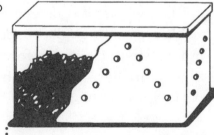

THE BEDDING: Tear up newspaper or corrugated

cardboard and dampen it with water. This bedding should always be moist, but not dripping wet. Loosely fill the bin about halfway with the moist bedding.

THE EARTHWORMS: Not just any worm will do! The best worms for a worm bin are red wigglers. You can get them at bait shops, or you can find them in compost heaps or at the bottom of a pile of leaves. They are about an inch and a half long, and are a reddish brown color. You will need about a half-pound of worms for a toaster-oven-size bin. Use fewer worms if your bin is smaller. These worms don't do well in really cold temperatures, so keep the worm bin in a place that stays above freezing.

WORM FOOD: Red wigglers are not fussy eaters. They will even eat their bedding! But they are vegetarians, and you should feed them about a cup of fresh fruit and vegetable scraps a week (make sure to bury the food in the bedding). They enjoy plain bread, pasta, grains, coffee grounds, and tea bags, too! They can eat their weight in food scraps each day. However, there are some foods that worms won't eat. They won't touch meat, bones, dairy products, or anything oily or covered in sauce.

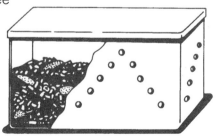

WORM BIN MAGIC: Do you have champion dirt-makers? After a month or so, you should know. It takes about that long for worms to make the vitamin-rich part of the soil (the other part is usually sand or ground-up rock).

OTHER THINGS TO TRY

The worm castings make great soil to grow plants in, so why not grow one? Put an old potato or avocado pit in the bin and watch it sprout. For another experiment, try placing two cups in the bin, one paper and the other plastic. What happens to them over time?

FUN FACTS

- Many people think that if you cut an earthworm in half, two new worms will grow from the pieces. However, this is not true. Many earthworms have been needlessly killed by people attempting to create two worms by cutting one in half. Do not try it!

- When you throw food scraps into the trash, they go to a landfill, and usually don't decompose for many years. Worms will decompose the scraps in just months.

ANIMAL TRACKS, ANIMAL DETECTIVE

So you want to be a detective? How about becoming an animal detective? There's no special training required except to keep your eyes on the ground to watch for feet. Not your feet! Animal feet! Well, to be more accurate, animal tracks. You can tell a lot about an animal by its feet and the tracks they leave behind. First, think about your own feet. If you saw prints that had five toes in a row and a long, oblong-shaped foot, it would be safe to guess that they were human prints (unless you lived near chimpanzees or gorillas, because their feet are very similar to ours). Suppose you saw a set of very small tracks with a mysterious line running between them. What clues would these tracks give you? Well, small tracks usually mean a small animal. The long line could mean a tail dragging. Could these tracks have been left by a mouse? You couldn't be sure, but that would be a good guess.

When you can't determine what the animal is simply by looking at its tracks, you have to look elsewhere for additional clues. For example, if you saw tracks leading to a tree and you saw a lot of teeth marks on the tree's bark, you could guess that the animal was eating the tree. Could it have been a deer or a beaver?

I-N-S-T-R-U-C-T-I-O-N-S

Look at the mystery scene on this page. Compare the animal tracks in the mystery scene to the animal track "mug shots." Can you name each animal in the mystery scene? Now write a story about what you think happened in the mystery scene.

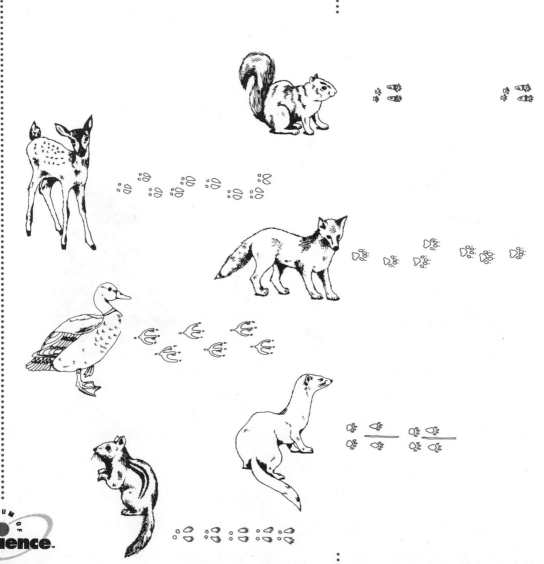

ACROSS

1 What has five hearts and no eyes, makes slime, and has a brain the size of a grain of sand?

6 Animals without backbones.

8 The vitamin-rich part of the soil is made by this animal.

9 Both insects and spiders belong to this large group of animals that have jointed legs.

12 Bats belong to the _____ family, which means "hand wing," because a bat's wing is really its hand.

13 A bird that lays eggs weighing three and a half pounds.

14 Animals, like people, that are covered with hair or fur. These are the only animals that drink milk.

15 Excreted by worms, it is a major component of soil.

16 Animals with backbones.

DOWN

1 A shell arthropods have on the outside of their bodies that provides structure and support.

2 Most scientific names are in this language.

3 They are all at least a little bigger than bacteria and can move around.

4 Birds whose feet are well designed for grabbing branches.

5 The Spanish word for bear.

7 Scientists who study animals.

10 An invertebrate with no shell that lives in the ocean and squirts black ink to protect itself.

11 Lions are good at this because their coats are yellow, just like the tall grasses in which they hide.

Solution Chapter 1 Crossword

Solution Chapter 2 Crossword

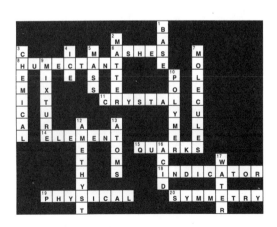

Solution Chapter 3 Crossword

Solution Chapter 4 Crossword

Solution Chapter 5 Crossword

Solution Chapter 6 Crossword

Solution Chapter 7 Crossword

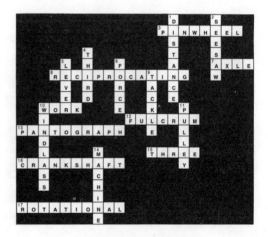

Solution Chapter 8 Crossword

Solution Chapter 9 Crossword

Solution Chapter 10 Crossword

Solution Chapter 11 Crossword

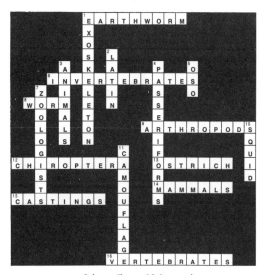

Solution Chapter 12 Crossword

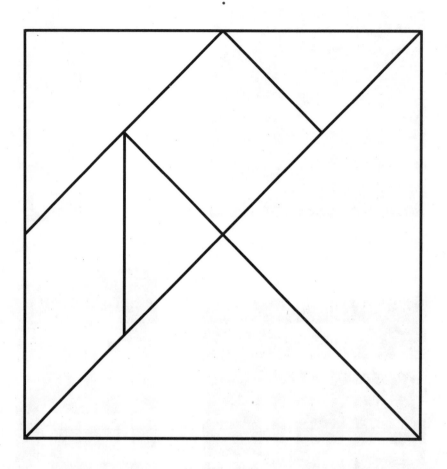

Tangram